SpringerBriefs in Mathematics

Series Editors

Nicola Bellomo
Michele Benzi
Palle Jorgensen
Tatsien Li
Roderick Melnik
Otmar Scherzer
Benjamin Steinberg
Lothar Reichel
Yuri Tschinkel
George Yin
Ping Zhang

T0212713

SpringerBriefs in Mathematics showcases expositions in all areas of mathematics and applied mathematics. Manuscripts presenting new results or a single new result in a classical field, new field, or an emerging topic, applications, or bridges between new results and already published works, are encouraged. The series is intended for mathematicians and applied mathematicians.

More information about this series at http://www.springer.com/series/10030

Elena Guardo • Adam Van Tuyl

Arithmetically Cohen-Macaulay Sets of Points in $\mathbb{P}^1 \times \mathbb{P}^1$

Springer

Elena Guardo
Dipartimento di Matematica e Informatica
University of Catania
Catania, Italy

Adam Van Tuyl
Department of Mathematics and Statistics
McMaster University
Hamilton, ON, Canada

ISSN 2191-8198 ISSN 2191-8201 (electronic)
SpringerBriefs in Mathematics
ISBN 978-3-319-24164-7 ISBN 978-3-319-24166-1 (eBook)
DOI 10.1007/978-3-319-24166-1

Library of Congress Control Number: 2015952874

Mathematics Subject Classification (2010): 13C14, 13H10, 14M05, 13A02, 13D02, 13D40, 41A05, 05A17

Springer Cham Heidelberg New York Dordrecht London
© The Authors 2015
This work is subject to copyright. All rights are reserved by the Publisher, whether the whole or part of the material is concerned, specifically the rights of translation, reprinting, reuse of illustrations, recitation, broadcasting, reproduction on microfilms or in any other physical way, and transmission or information storage and retrieval, electronic adaptation, computer software, or by similar or dissimilar methodology now known or hereafter developed.
The use of general descriptive names, registered names, trademarks, service marks, etc. in this publication does not imply, even in the absence of a specific statement, that such names are exempt from the relevant protective laws and regulations and therefore free for general use.
The publisher, the authors and the editors are safe to assume that the advice and information in this book are believed to be true and accurate at the date of publication. Neither the publisher nor the authors or the editors give a warranty, express or implied, with respect to the material contained herein or for any errors or omissions that may have been made.

Printed on acid-free paper

Springer International Publishing AG Switzerland is part of Springer Science+Business Media (www.springer.com)

For Mario, Roberto and Elisabetta
- EG

For Catherine and Iris
-AVT

Contents

Chapter 1
Introduction

The interpolation problem is a significant motivating problem in algebraic geometry and commutative algebra. Naively, the goal of the interpolation problem is to consider a collection of points in some ambient space, perhaps with some restrictions, and to describe all the polynomials that vanish at this collection. Introductions to this problem can be found in [7, 28, 62, 75]. The interpolation problem has applications to other areas of mathematics, including splines [30] and coding theory [40, 57].

One formulation of the interpolation problem is in terms of Hilbert functions of homogeneous ideals of finite sets of points in a projective space. More precisely, let $X = \{P_1,\ldots,P_s\}$ be a finite collection of distinct points in $\mathbb{P}^n$, an n-dimensional projective space. Associate with each point P_i a positive integer m_i, normally called its multiplicity. We would then like to determine the dimension of the vector space of homogeneous polynomials of each given degree that not only vanish on X, but also have the additional property that all their (m_i-1)-th partial derivatives also vanish at P_i. Algebraically, let $R = k[\mathbb{P}^n] = k[x_0,\ldots,x_n]$, and consider the homogeneous ideal $I(Z) := \bigcap_{i=1}^{s} I(P_i)^{m_i}$ of R, where $I(P_i)$ is the homogeneous ideal associated with P_i. The interpolation problem then asks what can be the Hilbert function of $R/I(Z)$, where the Hilbert function $H_Z(t) := \dim_k(R/I(Z))_t$ computes the dimension of the homogeneous component $(R/I(Z))_t$ of degree t of $R/I(Z)$ for all $t \in \mathbb{N}$. In particular, we would like an answer to the following problem.

Problem 1.1 (Interpolation Problem). *Classify the numerical functions $H : \mathbb{N} \to \mathbb{N}$ such that $H = H_Z$ is the Hilbert function of $R/I(Z)$ where $I(Z) = \bigcap_{i=1}^{s} I(P_i)^{m_i}$ for some set $X = \{P_1,\ldots,P_s\} \subseteq \mathbb{P}^n$ and multiplicities $m_i \geq 1$.*

When the set of points is reduced, that is, $m_i = 1$ for all i, we have a complete answer to the interpolation problem due to work of Geramita-Maroscia-Roberts [31] and Geramita-Gregory-Roberts [32]. The Hilbert functions of reduced finite sets of points in $\mathbb{P}^n$ are in one-to-one correspondence with the Hilbert functions of artinian quotients of $k[x_1,\ldots,x_n]$. Furthermore, by Macaulay's celebrated classification of Hilbert functions [68], we can explicitly determine which numerical functions are

© The Authors 2015

E. Guardo, A. Van Tuyl, *Arithmetically Cohen-Macaulay Sets of Points in* $\mathbb{P}^1 \times \mathbb{P}^1$, SpringerBriefs in Mathematics, DOI 10.1007/978-3-319-24166-1_1 the Hilbert functions of artinian quotients of $k[x_1,\dots,x_n]$. The situation, however, is less satisfactory when some of the multiplicities are greater than one. In this case we refer to the set of points as a collection of "fat points". The interpolation problem remains open even for the case of fat points in the projective plane; see [34, 35, 58] for more on this topic.

To study Problem 1.1, one is naturally lead to investigate the properties of the quotient ring $R/I(Z)$ associated to a set of (fat) points. In particular, the study of ideals of (fat) points in Problem 1.1 has motivated additional work beyond the study of Hilbert functions. For example, the graded minimal free resolutions of these ideals have also been studied (see, for example, the papers of Eisenbud and Popescu [25], Geramita and Maroscia [29], Hirschowitz and Simpson [64], and Lorenzini [67], among others). The book of Eisenbud [24] also contains a chapter on the graded minimal free resolutions of ideals of points in $\mathbb{P}^2$.

One way in which we can generalize Problem 1.1 is to remove the restriction that the elements of X be only points. For example, we can consider the situation where X is a union of linear varieties (points, lines, planes, etc.). Hartshorne and Hirschowitz's work [61] on the Hilbert functions of general lines in $\mathbb{P}^3$ is one of the first such examples of this idea. Although much less is known about this situation, the recent work of Ballico [4] and of Carlini-Catalisano-Geramita [13, 14] has given us some new insights into this problem.

Starting with the work of Giuffrida-Maggioni-Ragusa [36–38] in the 1990s, a new variation of the interpolation problem was introduced. Rather than studying points in a single projective space $\mathbb{P}^n$, one considers a set of points in $\mathbb{P}^{n_1}\times\cdots\times\mathbb{P}^{n_r}$, a multiprojective space[1]. By changing the ambient space, one usually finds that new and subtle arguments are needed to generalize results about points in $\mathbb{P}^n$ to the multiprojective setting since the two settings can be significantly different. For example, one of the main differences between points in $\mathbb{P}^n$ and points in $\mathbb{P}^{n_1}\times\cdots\times\mathbb{P}^{n_r}$ with $r\geq 2$ is that in the former, any collection of points (indeed, any zero-dimensional scheme) has a Cohen-Macaulay coordinate ring (see Definition 2.19), while in the latter, this is no longer true. Because the coordinate ring of a set of points in a multiprojective space is a multigraded ring, we can generalize the interpolation problem as follows:

Problem 1.2 (Interpolation Problem in $\mathbb{P}^{n_1}\times\cdots\times\mathbb{P}^{n_r}$). *Classify the numerical functions $H:\mathbb{N}^r\to\mathbb{N}$ such that $H=H_Z$ is the Hilbert function of a multigraded ring $R/I(Z)$ where $I(Z)=\bigcap_{i=1}^s I(P_i)^{m_i}$ for some set $X=\{P_1,\dots,P_s\}\subseteq\mathbb{P}^{n_1}\times\cdots\times\mathbb{P}^{n_r}$ and multiplicities $m_i\geq 1$.*

Attacks on the generalized interpolation problem are hampered by the fact that we know relatively little about the coordinate rings of sets of points in a multiprojective space when compared to a single projective space. The one exception, which is

[1]Note that because $\mathbb{P}^1\times\mathbb{P}^1$ (the case investigated by Giuffrida, *et al.*) is isomorphic to the quadric surface $\mathscr{Q}$ in $\mathbb{P}^3$, this variation can also be viewed as studying the interpolation problem on hypersurfaces; the paper of Huizenga [66] considers similar questions

the main focus of this monograph, is the case of arithmetically Cohen-Macaulay sets of points in $\mathbb{P}^1 \times \mathbb{P}^1$. As first shown by Giuffrida-Maggioni-Ragusa [36], Problem 1.2 can be answered for reduced arithmetically Cohen-Macaulay sets of points in $\mathbb{P}^1 \times \mathbb{P}^1$ because these sets of points enjoy additional geometric and combinatorial structure.

The goal of this monograph is to compile a comprehensive survey on the properties of arithmetically Cohen-Macaulay (ACM) sets of points in $\mathbb{P}^1 \times \mathbb{P}^1$, that is, those sets of points whose coordinate ring is Cohen-Macaulay. We have collected the various threads in the literature on ACM sets of points in $\mathbb{P}^1 \times \mathbb{P}^1$ with the aim of providing a self-contained and unified introduction to this topic. We will highlight the fact that ACM sets of points in $\mathbb{P}^1 \times \mathbb{P}^1$ are also very combinatorial in nature, i.e., many of the homological invariants of these points are encoded in numerical data that describes the configuration of points. Of particular note, we classify when a reduced set of points in $\mathbb{P}^1 \times \mathbb{P}^1$ is ACM (Theorem 4.11). We also present a complete solution to Problem 1.2 for reduced ACM sets of points in $\mathbb{P}^1 \times \mathbb{P}^1$ (Theorem 5.17). We also show how to use information about fat points in $\mathbb{P}^1 \times \mathbb{P}^1$ with an ACM support to reveal properties about fat lines in $\mathbb{P}^3$.

Our motivation to study points in a multiprojective setting is two-fold. First, although this monograph focuses on points in multiprojective spaces in order to understand Problem 1.2, it can also be seen as part of the much larger programme to understand the properties of multigraded rings. Multigraded rings appear naturally in commutative algebra and algebraic geometry, as for example, the coordinate ring of a blow-up and a Rees algebra. However, as highlighted in the paper of Cox-Dickenstein-Schenck [21], new techniques and methods are usually needed for the study of a multigraded situation. We note that understanding the Hilbert functions and the minimal free resolutions of the coordinate rings of points in multiprojective space is included among the list of open problems in commutative algebra found in the survey article of Peeva-Stillman [80]. Multigraded Hilbert functions have been studied, e.g., in the work of Aramova-Crona-De Negri and Crona [2, 22], but we are still far from a complete understanding of these functions; in particular, a multigraded version of Macaulay's theorem remains elusive contributing to the difficulty of answering Problem 1.2. Even determining a multigraded analog of the Castelnuovo-Mumford regularity, an important invariant that gives crude bounds on the size of a minimal free resolution, has proven to be quite subtle with a number of generalizations being proposed (see Hà [55], Maclagan-Smith [69], and Sidman-Van Tuyl [89]).

Second, besides enhancing our understanding of multigraded rings, studying sets of points in multiprojective spaces and their Hilbert functions (and other homological invariants) has applications to other areas of mathematics. Catalisano-Geramita-Gimigliano [16] proved that a specific value of the Hilbert function of a collection of fat points in a multiprojective space is related to a classical problem of algebraic geometry concerning the dimensions of certain secant varieties of Segre varieties. Chiantini-Sacchi [18] introduced the notion of a Segre function, which is defined in terms of particular values of the multigraded Hilbert function of a finite set of points, to deduce new results about tensors. Understanding the interpolation problem in complete generality may give us insight into these and other similar questions. In addition, the study of points in multiprojective space can be utilized to give new results about linear varieties in a projective space. By "forgetting" the multigraded structure, but keeping the graded structure, we can develop new results about unions of linear spaces (returning us to one of generalizations of the interpolation problem mentioned above). Specifically, by studying points in $\mathbb{P}^1 \times \mathbb{P}^1$, we can gain insight into special configurations of lines in $\mathbb{P}^3$. This fact will be exploited at the end of this monograph to apply our results to some open problems and conjectures.

We now summarize the contents of this monograph. In the next chapter (Chapter 2), we begin by reviewing the relevant constructions related to multiprojective spaces. As well, we provide a quick refresher on Cohen-Macaulay rings. Because our primary focus is on points in $\mathbb{P}^1 \times \mathbb{P}^1$, our discussion will centre on the bigraded ring $R = k[x_0, x_1, y_0, y_1]$. In Chapter 3, we turn our attention to the basic properties of points in $\mathbb{P}^1 \times \mathbb{P}^1$. In particular, we introduce a combinatorial description for these sets of points. As we will stress throughout this monograph, the combinatorics of the points in $\mathbb{P}^1 \times \mathbb{P}^1$, e.g., the number of points lying on a line, provide information about the algebraic invariants of the associated coordinate ring. We introduce the bigraded Hilbert function, and discuss some of the properties of the Hilbert functions of sets of points in $\mathbb{P}^1 \times \mathbb{P}^1$. As we show, the Hilbert function captures some of this combinatorial information. We end the chapter with an introduction to the notion of separators in the context of $\mathbb{P}^1 \times \mathbb{P}^1$.

The fourth chapter brings us to arithmetically Cohen-Macaulay sets of points, the principle object studied in this monograph. ACM sets of points in $\mathbb{P}^1 \times \mathbb{P}^1$ are sets of points whose associated bigraded coordinate rings are Cohen-Macaulay rings. As has been mentioned, sets of points in $\mathbb{P}^{n_1} \times \cdots \times \mathbb{P}^{n_r}$ may or may not be ACM. This fact motivates the need for one of the fundamental results of this monograph, namely, a classification of arithmetically Cohen-Macaulay sets of points in $\mathbb{P}^1 \times \mathbb{P}^1$ (see Theorem 4.11). One can characterize these points in many different ways: in terms of their Hilbert functions, in terms of two different combinatorial characterizations, in terms of the geometry of the points, and in terms of the notion of a separator. These results, which have appeared in the literature at different points in the last twenty years, are combined into one main result.

Because arithmetically Cohen-Macaulay sets of points in $\mathbb{P}^1 \times \mathbb{P}^1$ can be characterized through a combinatorial description, one can ask what other information can be gleaned from this description. We show how to compute the bigraded minimal free resolution, and consequently, the bigraded Hilbert function, directly from this numerical information. In Chapter 5, we carry out this programme. The results of this chapter will also answer the interpolation problem (Problem 1.2) for reduced ACM points in $\mathbb{P}^1 \times \mathbb{P}^1$ because we can completely characterize the Hilbert functions of ACM reduced sets of points in this context.

In Chapter 6, we turn our attention to fat points in $\mathbb{P}^1 \times \mathbb{P}^1$. Like the case of fat points in $\mathbb{P}^n$, less is known about fat points in $\mathbb{P}^1 \times \mathbb{P}^1$ when compared to their reduced cousins. Nevertheless, we can determine when a collection of fat points is arithmetically Cohen-Macaulay. Our main result in this chapter has similarities to the main result of Chapter 4: we can characterize when a set of fat points is arithmetically Cohen-Macaulay from either its Hilbert function or a combinatorial description of the points and multiplicities (see Theorem 6.21). As in the reduced case, for these ideals we can determine the bigraded Betti numbers in the bigraded minimal free resolution from this combinatorial data, and consequently, we can compute the Hilbert function. While we can compute the Hilbert functions of an ACM set of points in $\mathbb{P}^1 \times \mathbb{P}^1$, the interpolation problem for a set of ACM fat points in $\mathbb{P}^1 \times \mathbb{P}^1$, that is, classify what functions can be the Hilbert function of an ACM set of fat points in $\mathbb{P}^1 \times \mathbb{P}^1$, remains an open problem.

In Chapter 7, we consider the case of double points, i.e., each point has multiplicity two, in $\mathbb{P}^1 \times \mathbb{P}^1$. We further restrict to the case that the underlying support is arithmetically Cohen-Macaulay. While the results of Chapter 6 imply that these schemes are rarely arithmetically Cohen-Macaulay, we can still exploit the combinatorial description of the underlying set of points to compute the bigraded minimal free resolution of these double points. Our result is presented as an algorithm.

As noted above, if we only consider the coarse grading of a coordinate ring of a set of (fat) points in $\mathbb{P}^1 \times \mathbb{P}^1$, the resulting ring can also be viewed as the coordinate ring of a union of (fat) lines in $\mathbb{P}^3$. In Chapter 8, we show how to use the results of this monograph, most notably the results of Chapter 7, to answer questions and problems of interest in commutative algebra. In particular, our work gives additional evidence to a conjecture of Römer about the shifts that appear in graded free resolution of an ideal. We also give a negative answer to a question of Huneke on the properties of symbolic powers of ideals.

We end each chapter with a brief historical overview and cite the related results in the literature. We also include, where relevant, some open questions which we hope will inspire future research.

When preparing this manuscript, we have worked from the assumption that our readers are familiar with the basic material in Cox-Little-O'Shea [19] or Hassett [62]. In particular, the reader should be familiar with the algebra-geometry dictionary between graded rings and projective varieties. We also assume the reader has seen Hilbert functions and minimal free resolutions, e.g., see Eisenbud [24], Schenck [87] or Peeva [79]. Having said this, we have endeavoured to keep this text as self-contained as possible. While most of the results of this monograph have previously appeared, many of our proofs are new.

We would like to conclude this introduction with some personal comments. The study of points of multiprojective spaces is a topic that has been of great interest to both authors. We both "cut our teeth" on this subject while working on our PhD theses. Indeed, some of the results in this book first started out as results in our dissertations. After working on this topic with each other and our co-authors for almost fifteen years, we thought it would be a good idea to step back, and take some time to collate the material into a single volume.

Mathematical results do not normally arise in a vacuum. This fact is definitely true for the work that you are about to read. We would like to thank a number of people who have made this book possible. First, we would like to thank our supervisors, Alfio Ragusa and Tony Geramita, for having the foresight to pick an interesting, and mostly unexplored, area for us to hone our mathematical abilities. We also want to thank Juan Migliore, Tony, and Alfio for giving us the opportunity to collaborate for the first time at the PRAGMATIC conference of 2000 in Catania, Italy. It is doubtful if our many future collaborations, including this book, would exist without the wonderful opportunities provided at this conference. In addition, we would like to thank our co-authors, Chris Francisco, Tài Hà, Brian Harbourne, and Jessica Sidman, who worked on this topic with us over the years. We hope you learnt as much from us as we learnt from you. As well, we would like to thank Tony Geramita for providing some preliminary feedback on this manuscript and the anonymous referees for their valuable comments. It goes without saying, we take responsibility for any errors that remain (and please email us any corrections).

Finally, we acknowledge the financial support provided over the years by the Natural Sciences and Engineering Research Council of Canada (NSERC), and Istituto Nazionale di Alta Matematica (INdAM). We would also like to thank the Università di Catania and Lakehead University for helping to enable our many collaborations.

Elena Guardo
Catania
Adam Van Tuyl
Thunder Bay & Hamilton
Fall 2013–Summer 2015

Chapter 2
The biprojective space $\mathbb{P}^1 \times \mathbb{P}^1$

In this chapter we introduce bigraded rings and their properties, we introduce the biprojective space $\mathbb{P}^1 \times \mathbb{P}^1$ and its subvarieties, and we introduce the required background on Cohen-Macaulay rings. In particular, we build the correspondence between the bihomogeneous ideals of the bigraded ring $R = k[x_0,x_1,y_0,y_1]$ and the varieties of $\mathbb{P}^1 \times \mathbb{P}^1$, mimicking the well-known correspondence between graded ideals of a polynomial ring and the varieties of $\mathbb{P}^n$. While many of the results of this chapter extend quite naturally to any multiprojective space $\mathbb{P}^{n_1} \times \cdots \times \mathbb{P}^{n_r}$, we will focus primarily on the case of $\mathbb{P}^1 \times \mathbb{P}^1$ (see the discussion at the end of the chapter for what is known in the general setting). Similarly, our discussion of Cohen-Macaulay rings takes place within the context of the bigraded ring $R = k[x_0,x_1,y_0,y_1]$.

Throughout this monograph, k denotes an algebraically closed field of characteristic zero. In addition, $\mathbb{N} := \{0,1,2,\ldots\}$ denotes the set of non-negative integers. We let $\mathbb{N}^2 := \mathbb{N} \times \mathbb{N}$ and let $\preceq$ denote the natural partial order on the elements of $\mathbb{N}^2$ defined by $(a,b) \preceq (c,d)$ in $\mathbb{N}^2$ if and only if $a \leq c$ and $b \leq d$.

2.1 Bigraded rings

Let $R := k[x_0,x_1,y_0,y_1]$ be a polynomial ring with coefficients in k. We let $\mathbf{m} := (x_0,x_1,y_0,y_1)$. Set $\deg x_0 = \deg x_1 = (1,0)$ and $\deg y_0 = \deg y_1 = (0,1)$. A monomial $m = x_0^a x_1^b y_0^c y_1^d \in R$ has *bidegree* (or simply, *degree*) $\deg m = (a+b, c+d)$. We make the convention that 0, the additive identity, has $\deg 0 = (i,j)$ for all $(i,j) \in \mathbb{N}^2$. Note that the elements of k all have degree $(0,0)$.

For each $(i,j) \in \mathbb{N}^2$, let $R_{i,j}$ denote the vector space over k spanned by all the monomials of degree (i,j). (Because $0 \in R_{i,j}$ for all $(i,j) \in \mathbb{N}^2$, $R_{i,j}$ satisfies all the axioms of a vector space). The ring R is then a *bigraded ring* because there exists a direct sum decomposition

© The Authors 2015

E. Guardo, A. Van Tuyl, *Arithmetically Cohen-Macaulay Sets of Points in* $\mathbb{P}^1 \times \mathbb{P}^1$,
SpringerBriefs in Mathematics, DOI 10.1007/978-3-319-24166-1_2

$$R = \bigoplus_{(i,j)\in\mathbb{N}^2} R_{i,j}$$

such that $R_{i,j}R_{k,l} \subseteq R_{i+k,j+l}$ for all $(i,j),(k,l) \in \mathbb{N}^2$.

An element $F \in R$ is *bihomogeneous* if $F \in R_{i,j}$ for some $(i,j) \in \mathbb{N}^2$. If F is bihomogeneous, we say its *degree* is $\deg F = (i,j)$. Any polynomial $F \in R$ can be written uniquely as $F = F_1 + \cdots + F_t$ where each F_i is bihomogeneous. We call the F_i's the *bihomogeneous terms* of F.

Example 2.1. The element $F = x_0^2x_1y_0^3 + 17x_0x_1^2y_0^2y_1 - 34x_0^3y_1^3$ is a bihomogeneous element of R because it is an element of $R_{3,3}$. In particular, $\deg F = (3,3)$. On the other hand, the element $G = -16x_0x_1y_0y_1 + 14x_1^4$ is not bihomogeneous because $\deg(-16x_0x_1y_0y_1) = (2,2)$ but $\deg 14x_1^4 = (4,0)$. Note that if R has the standard grading , i.e., $\deg x_0 = \deg x_1 = \deg y_0 = \deg y_1 = 1$, then G is a homogeneous element of degree 4.

Suppose that $I = (F_1,\ldots,F_r) \subseteq R$ is an ideal. If each F_i is bihomogeneous, then we say that I is a *bihomogeneous ideal*. Just as in the standard graded case, it can be shown that I is a bihomogeneous ideal if and only if for every $F \in I$, all of the bihomogeneous terms of F also belong to I.

If $I \subseteq R$ is any ideal, then set $I_{i,j} := I \cap R_{i,j}$ for all $(i,j) \in \mathbb{N}^2$. Each $I_{i,j}$ is a subvector space of $R_{i,j}$, and furthermore, $I \supseteq \bigoplus_{(i,j)\in\mathbb{N}^2} I_{i,j}$. If I is bihomogeneous, then we have an equality, i.e., $I = \bigoplus_{(i,j)\in\mathbb{N}^2} I_{i,j}$, because the bihomogeneous terms of F belong to I if $F \in I$. When I is a bihomogeneous ideal of R, then the quotient ring R/I also inherits a bigraded ring structure. In particular, let $(R/I)_{i,j} := R_{i,j}/I_{i,j}$ for all $(i,j) \in \mathbb{N}^2$. Then

$$R/I = \bigoplus_{(i,j)\in\mathbb{N}^2} (R/I)_{i,j}.$$

An R-module M is a *bigraded R-module* if it has a direct sum decomposition

$$M = \bigoplus_{(i,j)\in\mathbb{N}^2} M_{i,j}$$

with the property that $R_{i,j}M_{k,l} \subseteq M_{i+k,j+l}$ for all $(i,j),(k,l) \in \mathbb{N}^2$. If I is a bihomogeneous ideal of R, then I and R/I are both examples of bigraded R-modules. Another example is the polynomial ring R but with a *shifted grading*. Specifically, let $(a,b) \in \mathbb{N}^2$. Then $R(-a,-b)$ is the polynomial ring, but the (i,j)-th graded piece of $R(-a,-b)$ is defined by

$$R(-a,-b)_{i,j} := R_{i-a,j-b}.$$

Note that $R_{i,j} = 0$ if $(0,0) \not\preceq (i,j)$.

Remark 2.2. Any bigraded R-module $M = \bigoplus_{(i,j)\in\mathbb{N}^2} M_{i,j}$ can also be viewed as a graded R-module if we view R as a standard graded ring, i.e., $\deg x_i = \deg y_i = 1$ for $i = 0,1$. Indeed, for each $t \in \mathbb{N}$, set

$$M_t := \bigoplus_{i+j=t} M_{i,j}.$$

Then $M = \bigoplus_{t\in\mathbb{N}} M_t$ is a graded R-module. So, for example, if I is a bihomogeneous ideal of $R = k[x_0,x_1,y_0,y_1]$, then we can also view R/I as a graded ring using the standard grading. We will find it expedient from time-to-time to "forget" the bigraded structures of our modules and consider only the standard graded structure.

Because of our interest in the interpolation problem, we shall be interested in the bigraded Hilbert function, and its associated first difference function, both of which are defined below.

Definition 2.3. Let I be a bihomogeneous ideal of $R = k[x_0,x_1,y_0,y_1]$. The *Hilbert function of R/I* is the numerical function $H_{R/I} : \mathbb{N}^2 \to \mathbb{N}$ defined by

$$H_{R/I}(i,j) := \dim_k(R/I)_{i,j} = \dim_k R_{i,j} - \dim_k I_{i,j}.$$

Example 2.4. As a simple example, we compute the Hilbert function of R/I when $I = (0)$. In this case, $H_{R/I}(i,j) = \dim_k R_{i,j}$. But we now have

$$\dim_k R_{i,j} = (i+1)(j+1) \text{ for all } (i,j) \in \mathbb{N}^2$$

because there are precisely $(i+1)(j+1)$ monomials of the form $x_0^a x_1^b y_0^c y_1^d$ with $a+b = i$ and $c+d = j$, and furthermore, these monomials form a basis for the k-vector space $R_{i,j}$.

Notation 2.5. When R/I is a bigraded ring, we write the output of the Hilbert function of R/I as an infinite matrix, where the initial row and column are indexed with 0 as opposed to 1. By Example 2.4, $\dim_k R_{i,j} = (i+1)(j+1)$ for all $(i,j) \in \mathbb{N}^2$. Consequently, the Hilbert function $H_{R/I}$ when $I = (0)$ is given by

$$H_{R/I} = \begin{bmatrix} 1 & 2 & 3 & 4 & 5 & \cdots \\ 2 & 4 & 6 & 8 & 10 & \cdots \\ 3 & 6 & 9 & 12 & 15 & \cdots \\ 4 & 8 & 12 & 16 & 20 & \cdots \\ 5 & 10 & 15 & 20 & 25 & \cdots \\ \vdots & \vdots & \vdots & \vdots & \vdots & \ddots \end{bmatrix}$$

Example 2.6. We consider a slightly more interesting example. Let

$$I = (x_0,y_0) \cap (x_1,y_1) = (x_0x_1, x_0y_1, x_1y_0, y_0y_1).$$

The ideal I is a monomial ideal and clearly bihomogeneous. Now $\dim_k I_{0,0} = \dim_k I_{1,0} = \dim_k I_{0,1} = 0$, which can be seen by the fact that the generators of I have degree $(2,0)$, $(1,1)$, or $(0,2)$. Now consider any $(i,j) \in \mathbb{N}^2 \setminus \{(0,0),(1,0),(0,1)\}$. Every monomial of degree (i,j) is an element of $I_{i,j}$ except $x_0^i y_0^j$ and $x_1^i y_1^j$, whence $\dim_k I_{i,j} = (i+1)(j+1) - 2$. So, the Hilbert function of R/I is given by

$$H_{R/I} = \begin{bmatrix} 1 & 2 & 2 & \cdots \\ 2 & 2 & 2 & \cdots \\ 2 & 2 & 2 & \cdots \\ \vdots & \vdots & \vdots & \ddots \end{bmatrix}.$$

In the standard graded case, information about R/I can be found in the first difference of its Hilbert function. An analog of the first difference Hilbert function can also be defined for bigraded Hilbert functions.

Definition 2.7. Let $H : \mathbb{N}^2 \to \mathbb{N}$ be a numerical function. The *first difference function of* H , denoted ΔH, is the function $\Delta H : \mathbb{N}^2 \to \mathbb{N}$ defined by

$$\Delta H(i,j) := H(i,j) - H(i-1,j) - H(i,j-1) + H(i-1,j-1)$$

where $H(i,j) = 0$ if $(i,j) \not\succeq (0,0)$.

Example 2.8. We computed the Hilbert function of R/I when $I = (x_0,y_0) \cap (x_1,y_1)$ in Example 2.6. The first difference function of $H_{R/I}$ is then given by

$$\Delta H_{R/I} = \begin{bmatrix} 1 & 1 & 0 & \cdots \\ 1 & -1 & 0 & \cdots \\ 0 & 0 & 0 & \cdots \\ \vdots & \vdots & \vdots & \ddots \end{bmatrix}.$$

An R-module homomorphism between two bigraded R-modules M and N, say $\varphi : M \to N$, has *degree* $(0,0)$ if $\varphi(M_{i,j}) \subseteq N_{i,j}$ for all $(i,j) \in \mathbb{N}^2$. Just as in the graded case, if

$$0 \to M_p \to M_{p-1} \to \cdots \to M_1 \to M_0 \to 0$$

is an exact sequence of bigraded R-modules with all morphisms with degree $(0,0)$, then we have the following relation among the Hilbert functions of $M_0,\ldots,M_p$:

$$H_{M_0}(i,j) - H_{M_1}(i,j) + H_{M_2}(i,j) - \cdots + (-1)^p H_{M_p}(i,j) = 0$$

for all $(i,j) \in \mathbb{N}^2$.

Associated with any bigraded R-module M is a *bigraded minimal free resolution*, that is, an exact sequence with degree $(0,0)$ homomorphisms of the form

$$0 \longrightarrow \bigoplus_{(j_1,j_2)\in\mathbb{N}^2} R(-j_1,-j_2)^{\beta_{l,(j_1,j_2)}(M)} \longrightarrow \cdots \longrightarrow$$

$$\bigoplus_{(j_1,j_2)\in\mathbb{N}^2} R(-j_1,-j_2)^{\beta_{1,(j_1,j_2)}(M)} \longrightarrow \bigoplus_{(j_1,j_2)\in\mathbb{N}^2} R(-j_1,-j_2)^{\beta_{0,(j_1,j_2)}(M)} \longrightarrow M \longrightarrow 0.$$

For details on how to construct this resolution, see [20, Chapter 6]; note that while this reference describes how to construct the graded minimal free resolution of a graded ideal, by taking into account the bigrading, we can adapt this procedure to construct a bigraded minimal free resolution. By Hilbert's Syzygy Theorem [20, Theorem 2.1], $l \leq 4$ because we are considering ideals in a polynomial ring with four variables. The numbers $\beta_{i,(j_1,j_2)}(M)$ are the *bigraded Betti numbers* of M.

2.2 Biprojective space $\mathbb{P}^1 \times \mathbb{P}^1$

We now generalize the definition of a projective space $\mathbb{P}^n$ and its subvarieties to a multiprojective setting. While multiprojective spaces can be defined using the modern scheme language, it will suffice for our purposes to only consider the classical construction, i.e., the algebra-geometry dictionary between radical ideals and closed subsets. We only sketch out the relevant details to describe the algebra-geometry dictionary for bihomogeneous ideals of $R = k[x_0,x_1,y_0,y_1]$ and closed subsets of $\mathbb{P}^1 \times \mathbb{P}^1$. The proofs are similar to the graded case (see, for example, the book of Cox, Little, and O'Shea [19]) so they are omitted.

The *biprojective space* $\mathbb{P}^1 \times \mathbb{P}^1$ is the set of equivalence classes

$$\mathbb{P}^1 \times \mathbb{P}^1 := \left\{ ((a_0,a_1),(b_0,b_1)) \in k^2 \times k^2 \;\middle|\; \begin{array}{c} \text{neither } (a_0,a_1) \neq (0,0) \\ \text{nor } (b_0,b_1) \neq (0,0) \end{array} \right\}_{/\sim}$$

where $\sim$ is the equivalence relation $((a_0,a_1),(b_0,b_1)) \sim ((c_0,c_1),(d_0,d_1))$ if there exist nonzero $\lambda,\mu \in k$ such that $(a_0,a_1) = (\lambda c_0, \lambda c_1)$ and $(b_0,b_1) = (\mu d_0, \mu d_1)$. An element of $\mathbb{P}^1 \times \mathbb{P}^1$ is called a *point*. We denote the equivalence class of $((a_0,a_1),(b_0,b_1))$ by $[a_0:a_1] \times [b_0:b_1]$. It follows that $[a_0:a_1]$, respectively $[b_0:b_1]$, is a point of $\mathbb{P}^1$.

If $F \in R$ is a bihomogeneous element of degree (i,j) and $P = [a_0:a_1] \times [b_0:b_1]$ is a point of $\mathbb{P}^1 \times \mathbb{P}^1$, then

$$F(\lambda a_0, \lambda a_1, \mu b_0, \mu b_1) = \lambda^i \mu^j F(a_0,a_1,b_0,b_1) \text{ for all nonzero } \lambda,\mu \in k.$$

To say that F vanishes at a point of $\mathbb{P}^1 \times \mathbb{P}^1$ is, therefore, a well-defined notion.

If T is any set of bihomogeneous elements of R, then we define

$$V(T) := \{P \in \mathbb{P}^1 \times \mathbb{P}^1 \mid F(P) = 0 \text{ for all } F \in T\}.$$

If I is a bihomogeneous ideal of R, then $V(I) := V(T)$ where T is the set of all bihomogeneous elements of I. If $I = (F_1, \dots, F_r)$, then $V(I) = V(\{F_1, \dots, F_r\})$.

The biprojective space $\mathbb{P}^1 \times \mathbb{P}^1$ can be endowed with a topology by defining the *closed sets* to be all the subsets of $\mathbb{P}^1 \times \mathbb{P}^1$ of the form $V(T)$ where T is a collection of bihomogeneous elements of R. If Y is a subset of $\mathbb{P}^1 \times \mathbb{P}^1$ that is closed and irreducible with respect to this topology, then we say Y is a *biprojective variety*, or simply, a *variety*.

If Y is any subset of $\mathbb{P}^1 \times \mathbb{P}^1$, then we set

$$I(Y) := \{F \in R \mid F(P) = 0 \text{ for all } P \in Y\}.$$

The set $I(Y)$ is a bihomogeneous ideal of R and is called the *bihomogeneous ideal associated with* Y, or simply, the *ideal associated with* Y. If $P \in \mathbb{P}^1 \times \mathbb{P}^1$ is a point, then we abuse notation and write $I(P)$ instead of $I(\{P\})$. We call $R/I(Y)$ the *bihomogeneous coordinate ring of* Y, or simply, the *coordinate ring of* Y. If $H_{R/I(Y)}$ is the Hilbert function of $R/I(Y)$, then we sometimes write H_Y for $H_{R/I(Y)}$, and we say H_Y is the *Hilbert function of* Y.

By adapting the proofs from the well-known homogeneous case, one can prove the following facts.

Theorem 2.9. *(i) If $I_1 \subseteq I_2$ are bihomogeneous ideals, then $V(I_1) \supseteq V(I_2)$.*
(ii) If $Y_1 \subseteq Y_2$ are subsets of $\mathbb{P}^1 \times \mathbb{P}^1$, then $I(Y_1) \supseteq I(Y_2)$.
(iii) For any two subsets Y_1, Y_2 of $\mathbb{P}^1 \times \mathbb{P}^1$, $I(Y_1 \cup Y_2) = I(Y_1) \cap I(Y_2)$.

Example 2.10. Consider the ideal of Example 2.6, that is, $I = (x_0, y_0) \cap (x_1, y_1)$. Since $V((x_0, y_0)) = \{[0:1] \times [0:1]\}$ and $V((x_1, y_1)) = \{[1:0] \times [1:0]\}$, we have

$$V(I) = \{[1:0] \times [1:0], [0:1] \times [0:1]\} \subseteq \mathbb{P}^1 \times \mathbb{P}^1.$$

One can build an algebra-geometry dictionary between bihomogeneous ideals of R and subvarieties of $\mathbb{P}^1 \times \mathbb{P}^1$ similar to the standard algebra-geometry dictionary between graded ideals and varieties in $\mathbb{P}^n$. The correspondence between these objects requires a bigraded version of the Nullstellensatz. Again, the proof follows as in the graded case, so it is omitted.

Theorem 2.11 (Bigraded Nullstellensatz). *If $I \subseteq R = k[x_0, x_1, y_0, y_1]$ is a bihomogeneous ideal and if $F \in R$ is a bihomogeneous polynomial with $\deg F \neq (0,0)$ such that $F(P) = 0$ for all $P \in V(I) \subseteq \mathbb{P}^1 \times \mathbb{P}^1$, then $F^t \in I$ for some $t > 0$.*

One difference between the standard graded case and the bigraded case is the notion of irrelevant ideals.

Definition 2.12. A bihomogeneous ideal I of $R = k[x_0, x_1, y_0, y_1]$ is called *projectively irrelevant* if either $(x_0, x_1)^t \subseteq I$ or $(y_0, y_1)^t \subseteq I$ for some integer t. An ideal $I \subseteq R$ is *projectively relevant* if it is not projectively irrelevant.

Example 2.13. To understand the nomenclature, suppose that I is an ideal with $(x_0,x_1)^t \subseteq I$. In particular, x_0^t and $x_1^t \in I$. If we now consider $V(I)$, we must have $V(I) = \emptyset$ since there is no point $P = [a:b] \times [c:d]$ that vanishes at both x_0^t and x_1^t. If P did vanish, we must have $a = b = 0$, which is not allowed.

The following result can be proved by adapting the proof of the graded case and using the Bigraded Nullstellensatz (Theorem 2.11).

Theorem 2.14. *There is a bijective correspondence between the non-empty closed subsets of $\mathbb{P}^1 \times \mathbb{P}^1$ and the bihomogeneous ideals of R that are radical, i.e., $I = \sqrt{I}$, and projectively relevant. The correspondence is given by $Y \mapsto I(Y)$ and $I \mapsto V(I)$.*

We conclude this section with a special case of Bezout's Theorem for curves in $\mathbb{P}^1 \times \mathbb{P}^1$.

Theorem 2.15 (Bigraded Bezout (Special Case)). *Let $F \in R = k[x_0,x_1,y_0,y_1]$ be a bihomogeneous form with $\deg F = (a,b)$, and let $H \in R_{1,0}$. If the curves $V(F)$ and $V(H)$ meet at more than b points (counting multiplicities), then $F = HF'$.*

Proof. After a change of coordinates, we can assume that $H = x_0$. After applying the Division Algorithm for polynomials (see [19, Theorem 2.3.3]), we have $F = F'x_0 + F''$ where $F'' = x_1^a G(y_0,y_1)$ and $G(y_0,y_1)$ is a homogeneous polynomial of degree b in the y_is.

Any point $P \in V(F) \cap V(H)$ must have the form $P = [0:1] \times [b_1 : b_2]$ because P lies on $V(H)$. But then $[b_1 : b_2] \in \mathbb{P}^1$ is a point that must vanish on $G(y_0,y_1)$. Because $G(y_0,y_1)$ is homogeneous polynomial of degree b, it has exactly b roots (counting multiplicity). So, if there are more than b points (counting multiplicities) that lie in the intersection, we must have $G(y_0,y_1) = 0$, as desired.

Although we will not require it in this monograph, here is the more general statement of Bezout's Theorem for the bigraded ring $R = k[x_0,x_1,y_0,y_1]$.

Theorem 2.16 (Bigraded Bezout). *Let $F,G \in R = k[x_0,x_1,y_0,y_1]$ be a bihomogeneous form with $\deg F = (a,b)$ and $\deg G = (c,d)$, and furthermore, suppose that G is irreducible. If the curves $V(F)$ and $V(G)$ meet at more than $ad + bc$ points (counting multiplicities), then $F = GF'$.*

Theorem 2.15 is just Theorem 2.16 when G is the irreducible form $H \in R_{1,0}$. Note that a result similar to Theorem 2.15 holds if we take an irreducible form $V \in R_{0,1}$. We leave it to the reader to verify the details.

2.3 Cohen-Macaulay rings

In this section, we review the relevant background on Cohen-Macaulay rings. We continue to work within the context of the bigraded ring $R = k[x_0,x_1,y_0,y_1]$, although these definitions and results hold more generally.

Definition 2.17. If $I \subseteq R$ is a bihomogeneous ideal, then a sequence $F_1, \ldots, F_r$ of elements is a *regular sequence modulo I* if and only if

(i) $(I, F_1, \ldots, F_r) \subseteq \mathbf{m}$,
(ii) $\overline{F}_1$ is not a zero-divisor in R/I, and
(iii) $\overline{F}_i$ is not a zero-divisor in $R/(I, F_1, \ldots, F_{i-1})$ for $1 < i \leq r$.

The *depth of R/I*, denoted $\mathrm{depth}(R/I)$, is the length of the maximum regular sequence modulo I.

Definition 2.18. If $I \subseteq R$ is a bihomogeneous ideal, then the *height* of a prime ideal $\wp$ in R/I, denoted $\mathrm{ht}_{R/I}(\wp)$, is the largest integer t such that there exist prime ideals $\wp_i$ of R/I such that $\wp_0 \subsetneq \wp_1 \subsetneq \cdots \subsetneq \wp_{t-1} \subsetneq \wp_t = \wp$. For any ideal I of R, the *Krull dimension of R/I*, denoted $\mathrm{K\text{-}dim}(R/I)$, is

$$\mathrm{K\text{-}dim}(R/I) := \sup\{\mathrm{ht}_{R/I}(\wp) \mid \wp \text{ a prime ideal of } R/I\}.$$

One always has the inequality $\mathrm{depth}(R/I) \leq \mathrm{K\text{-}dim}(R/I)$ (see, [88, Corollary 16.30]). The equality of these two invariants leads to an important class of rings.

Definition 2.19. If $I \subseteq R$ is a bihomogeneous ideal, then the ring R/I is *Cohen-Macaulay* if $\mathrm{depth}(R/I) = \mathrm{K\text{-}dim}(R/I)$.

In this monograph, we are primary interested in arithmetically Cohen-Macaulay varieties of $\mathbb{P}^1 \times \mathbb{P}^1$.

Definition 2.20. If $I = I(Y)$ is the bihomogeneous ideal of the variety $Y \subseteq \mathbb{P}^1 \times \mathbb{P}^1$, then we say Y is *arithmetically Cohen-Macaulay* (ACM) if $R/I(Y)$ is a Cohen-Macaulay.

The following results from homological algebra allow us to link the depth of a ring to its projective dimension.

Definition 2.21. The *projective dimension* of an R-module M, denoted $\mathrm{proj\text{-}dim}(M)$, is the length of a minimal free resolution of M.

Given a homogeneous ideal I of the ring R (not necessarily a bihomogeneous ideal), a beautiful result of Auslander and Buchsbaum links the projective dimension of R/I to the depth of R/I. We continue to assume that $R = k[x_0, x_1, y_0, y_1]$, although this result holds much, much more generally.

Theorem 2.22 (Auslander-Buchsbaum Formula). *Let I be a homogeneous ideal in the ring $R = k[x_0, x_1, y_0, y_1]$. Then*

$$\mathrm{proj\text{-}dim}(R/I) + \mathrm{depth}(R/I) = \mathrm{K\text{-}dim}(R) = 4.$$

Proof. See [79, 15.3] for both the general statement and its proof.

We can use short exact sequences to place bounds on the projective dimensions of R-modules. The following results can deduced from the Depth Lemma [98, Lemma 1.3.9] and the Auslander-Buchsbaum Formula.

Lemma 2.23. *Let M_1, M_2, and M_3 be R-modules, and suppose that we have a short exact sequence*

$$0 \to M_1 \to M_2 \to M_3 \to 0.$$

Then

(*i*) $\text{proj-dim}(M_3) \leq \max\{\text{proj-dim}(M_2), \text{proj-dim}(M_1)+1\}$.
(*ii*) $\text{proj-dim}(M_2) \leq \max\{\text{proj-dim}(M_1), \text{proj-dim}(M_3)\}$.

Two special classes of Cohen-Macaulay rings that we will encounter are complete intersections and artinian rings. We recall their definitions and properties.

Definition 2.24. An ideal $I \subseteq R$ is a *complete intersection* if it is generated by a regular sequence.

A complete intersection is also Cohen-Macaulay because of the following result.

Lemma 2.25. *Suppose that $\{F_1,\ldots,F_s\}$ is a regular sequence of homogeneous elements in the polynomial ring $R = k[x_0,\ldots,x_n]$. If I is the complete intersection generated by $\{F_1,\ldots,F_s\}$, then R/I is Cohen-Macaulay and* $\dim R/I = (n+1)-s$.

Proof. See [98, Lemma 1.3.10] and [98, Proposition 1.3.22]. Note that for both of these references, the ring is assumed to be a local ring. However, the results also hold for graded ideals in a polynomial ring.

The following lemma provides an explicit description of the bigraded minimal free resolution of a complete intersection generated by two bigraded forms in R. We shall find this a very useful tool.

Lemma 2.26. *Suppose that $I = (F,G) \subseteq R = k[x_0,x_1,y_0,y_1]$ is a complete intersection. Furthermore, suppose that* $\deg F = (a_1,a_2)$ *and* $\deg G = (b_1,b_2)$. *Then the bigraded minimal free resolution of I is given by*

$$0 \longrightarrow R(-a_1-b_1,-a_2-b_2) \xrightarrow{\phi_2} R(-a_1,-a_2) \oplus R(-b_1,-b_2) \xrightarrow{\phi_1} I \longrightarrow 0$$

where $\phi_1 = [G \ \ F]$ *and* $\phi_2 = \begin{bmatrix} F \\ -G \end{bmatrix}$.

Proof. This result is simply a special case of the fact that the Koszul complex is a minimal free resolution of a complete intersection (see [79, Theorem 14.7]). We also take into account the fact that the generators of I are bigraded.

Definition 2.27. Let $S = k[x_1,y_1]$ be a bigraded ring with $\deg x_1 = (1,0)$ and $\deg y_1 = (0,1)$. A bihomogeneous ideal $J \subseteq S$ is an *artinian ideal* if $\sqrt{J} = (x_1,y_1)$. The ring S/J is a *bigraded artinian quotient* if J is a bihomogeneous artinian ideal.

Remark 2.28. Because the dimension of an artinian ring is zero, and because we always have $0 \leq \text{depth}(S) \leq \text{K-dim}(S)$ when S is a quotient of a polynomial ring, we deduce that an artinian ring is Cohen-Macaulay.

Although we do not have a complete classification of bigraded Hilbert functions, we can classify the bigraded Hilbert functions of artinian quotients of $k[x_1, y_1]$.

Theorem 2.29. *Let $H : \mathbb{N}^2 \to \mathbb{N}$ be a numerical function. Then H is the Hilbert function of a bigraded artinian quotient of $k[x_1, y_1]$ if and only if*

(*i*) $H(0,0) = 1$,
(*ii*) $H(i,j) = 0$ *or* 1 *for all* $(i,j) \in \mathbb{N}^2$,
(*iii*) $H(i,j) = 1$ *for only finitely many* $(i,j) \in \mathbb{N}^2$, *and*
(*iv*) $H(i,j) = 0$ *implies that* $H(k,l) = 0$ *for all* $(i,j) \preceq (k,l) \in \mathbb{N}^2$.

Proof. ($\Rightarrow$) Suppose that there is a bihomogeneous artinian ideal $J \subseteq S = k[x_1, y_1]$ such that $H = H_{S/J}$. Because J is artinian, $J \neq (1)$, so $\dim_k(S/J)_{0,0} = 1$, that is, $H_{S/J}(0,0) = 1$, thus proving (i).

Because $\dim_k S_{i,j} = 1$ for all $(i,j) \in \mathbb{N}^2$, we have

$$H(i,j) = H_{S/J}(i,j) = \dim_k S_{i,j} - \dim_k J_{i,j} = 1 - \dim_k J_{i,j} \quad \text{for all } (i,j) \in \mathbb{N}^2.$$

Thus $H_{S/J}(i,j) = 0$ or 1 for all $(i,j) \in \mathbb{N}^2$, proving (ii).

Because J is artinian, there exist positive integers a and b such that $x_1^a \in J$ and $y_1^b \in J$. Thus, for all $(i,j) \in \mathbb{N}^2$, if $i \geq a$ or if $j \geq b$, then $\dim_k J_{i,j} = 1$, whence $H_{S/J}(i,j) = 0$. So, if $H_{S/J}(i,j) = 1$, then $(i,j) \preceq (a-1, b-1)$. Because there are only a finite number of such (i,j), this proves (iii).

Finally, if $H_{S/J}(i,j) = 0$, this means that $\dim_k J_{i,j} = 1$, and thus $\dim_k J_{k,l} = 1$ for all $(i,j) \preceq (k,l)$. This, in turn, implies that $H_{S/J}(k,l) = 0$, thus proving (iv).

($\Leftarrow$) Let H be a numerical function that satisfies (i) through (iv). In $S = k[x_1, y_1]$, let J be the ideal generated by $\{x_1^i y_1^j \mid H(i,j) = 0\}$. We claim that $H = H_{S/J}$ and J is artinian.

First, note that $\dim_k J_{i,j} = 1$ if and only if $H(i,j) = 0$. One direction of this statement follows directly from the definition of J. On the other hand, if $\dim_k J_{i,j} = 1$, then $x_1^i y_1^j \in J_{i,j}$. Thus, there is some generator $x_1^k y_1^l \in J_{k,l}$ with $(k,l) \preceq (i,j)$ that divides $x_1^i y_1^j$. Because $x_1^k y_1^l$ is a generator of J, $H(k,l) = 0$. But then by (iv), $H(i,j) = 0$ since $(k,l) \preceq (i,j)$. Note that $J \neq (1)$ since $H(0,0) = 1$ by (i). Finally, since $H(i,j) = 1$ for only a finite number of (i,j), there must exist an (i,j) of the form $(a,0)$ such that $H(a,0) = 0$. But this means $x_1^a \in J$. Similarly, we can find some $(0,b)$ such that $H(0,b) = 0$, which means $y_1^b \in J$. So, J is artinian.

2.4 Additional notes

The material in this chapter is quite standard. The first two sections are a natural generalization of the algebra-geometry dictionary from the graded case to the multigraded case. While we have focused on the case of $\mathbb{P}^1 \times \mathbb{P}^1$, everything we have presented in this chapter extends naturally to the case of $\mathbb{P}^{n_1} \times \cdots \times \mathbb{P}^{n_r}$. To the best of our knowledge, Van der Waerden [90] was the first to consider the multigraded version of the algebra-geometry dictionary (although he focused on the bigraded case, and left the general situation to the reader). Readers may also be interested in a later paper of Van der Waerden [91]. The construction of the multigraded dictionary can also be found in a paper of the second author [93]. A different approach using the language of schemes and the notion of Proj can be found in the PhD thesis of Vidal [97]. The results on Cohen-Macaulay rings can be found, in much greater generality, in most graduate level books on commutative algebra, e.g., [12, 24, 88].

As mentioned in Chapter 1, classifying bigraded (or multigraded) Hilbert functions remains an open problem. Ideally, one would like to find a multigraded analog of Macaulay's classification [68] of Hilbert functions of graded quotients of $k[x_1,\ldots,x_n]$. Van der Waerden's work [90] contains some early results on this problem. More recently, Aramova-Crona-De Negri [2] and Crona [22] presented some necessary conditions on the growth of bigraded Hilbert functions. The case of the bigraded ring $k[x_1,y_1,\ldots,y_n]$ with $\deg x_1 = (1,0)$ and $\deg y_i = (0,1)$ with $i = 1,\ldots,n$ was studied by the second author [92]. In this chapter, we have only presented this result for the case of $k[x_1,y_1]$. Any improvements on these results would be a welcome addition to the literature.

Chapter 3
Points in $\mathbb{P}^1 \times \mathbb{P}^1$

In this chapter we develop the basic properties of sets of points in $\mathbb{P}^1 \times \mathbb{P}^1$. We begin by describing the bihomogeneous ideal $I(P)$ associated with a point $P \in \mathbb{P}^1 \times \mathbb{P}^1$. We also discuss the connection between sets of points X in $\mathbb{P}^1 \times \mathbb{P}^1$ and special configurations of lines in $\mathbb{P}^3$. Because $\mathbb{P}^1 \times \mathbb{P}^1$ is isomorphic to the ruled quadric surface in $\mathbb{P}^3$, we can visualize a collection of points X as sitting within a grid of lines. After describing how to represent a set of points X in $\mathbb{P}^1 \times \mathbb{P}^1$, we extract some combinatorial information from this representation. This combinatorial data is linked to some algebraic invariants of $R/I(X)$. In particular, we show how to compute all but a finite number of values of the Hilbert function of $X \subseteq \mathbb{P}^1 \times \mathbb{P}^1$ directly from this combinatorial description. We conclude the chapter with an introduction to separators of points. Roughly speaking, a separator is a tool to study how the algebraic invariants of X and $X \setminus \{P\}$, with $P \in X$, are related.

3.1 Points

A point $P \in \mathbb{P}^1 \times \mathbb{P}^1$ has the form $P = A \times B$ where $A, B \in \mathbb{P}^1$. Note that the points A and B need not be distinct. We adopt the convention that if we are only interested in the point, but not its distinct coordinates, then we use P, but if we wish to keep track of our coordinates, then we use As and Bs. Given a point $P = A \times B$, its associated bihomogeneous ideal is given by

$$I(P) = \{F \in R \mid F(P) = 0\} \subseteq R = k[x_0, x_1, y_0, y_1].$$

Our first result gives some information about the ideal $I(P)$.

Theorem 3.1. *Let $I(P)$ be the bihomogeneous ideal in the bigraded ring $R = k[x_0, x_1, y_0, y_1]$ associated with a point $P \in \mathbb{P}^1 \times \mathbb{P}^1$. Then*

© The Authors 2015

E. Guardo, A. Van Tuyl, *Arithmetically Cohen-Macaulay Sets of Points in $\mathbb{P}^1 \times \mathbb{P}^1$*,
SpringerBriefs in Mathematics, DOI 10.1007/978-3-319-24166-1_3

(*i*) *$I(P)$ is a prime ideal of R.*
(*ii*) *$I(P) = (H,V)$ where $\deg H = (1,0)$ and $\deg V = (0,1)$.*
(*iii*) *Let $X = \{P_1,\ldots,P_s\} \subseteq \mathbb{P}^1 \times \mathbb{P}^1$ be a set of s distinct points and suppose that $I(P_i)$ is the ideal associated with the point P_i. Then $I(X) = I(P_1) \cap I(P_2) \cap \cdots \cap I(P_s)$.*

Proof. (*i*) If $FG \in I(P)$, then $(FG)(P) = F(P)G(P) = 0$. Hence, either F or G must vanish at P, and thus is an element of $I(P)$.

(*ii*) Suppose that $P = [a_0 : a_1] \times [b_0 : b_1]$. Assume for the moment that $a_1 \neq 0$ and $b_1 \neq 0$. Thus, we can assume that $P = [a_0 : 1] \times [b_0 : 1]$. Set

$$I = (x_0 - a_0x_1, y_0 - b_0y_1).$$

Then $I \subseteq I(P)$ because all of the generators of I vanish at P. It suffices to show that $I(P) = I$ because then we can take $H = x_0 - a_0x_1$ and $V = y_0 - b_0y_1$.

To show the reverse inclusion $I(P) \subseteq I$, consider any $F \in I(P)$. Because $I(P)$ is a bihomogeneous ideal, we can assume that F is a bihomogeneous polynomial of degree (i,j). Let $>$ be any monomial ordering of R with $x_0 > x_1 > y_0 > y_1$. With this monomial order, apply the Division Algorithm [19, Theorem 2.3.3] to F using the generators of I to get

$$F = F_1(x_0 - a_0x_1) + F_2(y_0 - b_0y_1) + F_3$$

where $\deg F_1 = (i-1,j)$, $\deg F_2 = (i,j-1)$, and $\deg F_3 = (i,j)$. Furthermore, by the Division Algorithm, we know that neither leading term of $x_0 - a_0x_1$ nor $y_0 - b_0y_1$, that is, x_0 and y_0, respectively, divides any monomial term in F_3. So, only the variables x_1 and y_1 appear in F_3. Because there is exactly one monomial of degree (i,j) in these two variables, namely, $x_1^iy_1^j$, this implies that $F_3 = cx_1^iy_1^j$ for some constant $c \in k$. But because $F \in I(P)$, when we evaluate $F(x_0,x_1,y_0,y_1)$ at the point $P = [a_0 : 1] \times [b_0 : 1]$, we get

$$0 = F(a_0,1,b_0,1) = F_1(a_0,1,b_0,1)(a_0 - a_0 \cdot 1) + F_2(a_0,1,b_0,1)(b_0 - b_0 \cdot 1) + c1^i1^j = c.$$

In other words, $F = F_1(x_0 - a_0x_1) + F_2(y_0 - b_0y_1) \in I$.

Note that if $a_1 = 0$, then $a_0 \neq 0$. We can repeat the above argument, but use a monomial ordering with $x_1 > x_0$ to find the required forms. A similar argument holds if $b_1 = 0$.

(*iii*) This fact follows from Theorem 2.9 (*iii*).

In the proof of Theorem 3.1 (*ii*) we actually proved the following fact.

Corollary 3.2. *Let $P = A \times B \in \mathbb{P}^1 \times \mathbb{P}^1$. If $A = [a_0 : a_1] \in \mathbb{P}^1$ and $B = [b_0 : b_1] \in \mathbb{P}^1$, then $I(P) = (a_1x_0 - a_0x_1, b_1y_0 - b_0y_1)$.*

Remark 3.3. The coordinate rings of $\mathbb{P}^3$ and $\mathbb{P}^1 \times \mathbb{P}^1$ are the same rings, but with different gradings. Thus, if $P \in \mathbb{P}^1 \times \mathbb{P}^1$ is a point with associated bihomogeneous ideal $I(P)$, when we consider $I(P)$ as a graded ideal of $R = k[x_0,x_1,y_0,y_1]$, then $I(P)$

corresponds geometrically to a line in $\mathbb{P}^3$. Moreover, for any collection of points $X = \{P_1,\dots,P_s\} \subseteq \mathbb{P}^1 \times \mathbb{P}^1$, if we only consider the standard graded ideal $I(X)$, then $I(X)$ corresponds to a union of lines in $\mathbb{P}^3$.

In fact, $I(X)$ corresponds to a special configuration of lines, which we now describe. Let L_1 be the line in $\mathbb{P}^3$ associated to the ideal (y_0,y_1), and let L_2 be the line associated to (x_0,x_1). Note that L_1 and L_2 are two skew lines in $\mathbb{P}^3$, that is, $L_1 \cap L_2 = \emptyset$.

Each point $P_i \in X \subseteq \mathbb{P}^1 \times \mathbb{P}^1$ can be written as $P_i = [a_{i,0} : a_{i,1}] \times [b_{i,0} : b_{i,1}]$. By Corollary 3.2, $I(P_i) = (a_{i,1}x_0 - a_{i,0}x_1, b_{i,1}y_0 - b_{i,0}y_1)$. As a line in $\mathbb{P}^3$, the line defined by $I(P_i)$ intersects L_1 at $[a_{i,0} : a_{i,1} : 0 : 0]$ and L_2 at $[0 : 0 : b_{i,0} : b_{i,1}]$. So, as a graded ideal, $I(X)$ defines a union of lines in $\mathbb{P}^3$ where each line intersects both L_1 and L_2, and furthermore, the coordinates of each P_i describe where the line defined by $I(P_i)$ intersects these two skew lines.

It follows that our study of points in $\mathbb{P}^1 \times \mathbb{P}^1$ can be seen as an investigation of these special unions of lines in $\mathbb{P}^3$. As we show in the sequel, by taking advantage of the bigraded structure, we can deduce some interesting results about these varieties (see, for example, the results of Chapter 8).

Our focus until Chapter 5 shall be on reduced sets of points.

Definition 3.4. A set of points $X \subseteq \mathbb{P}^1 \times \mathbb{P}^1$ is *reduced* if $I(X) = \sqrt{I(X)}$.

When X consists of distinct points, then X is reduced. Non-reduced sets of points, sometimes called fat points, are introduced in Chapter 6.

The following lemma gives us some useful algebraic information; note that the proof can also be adapted to show the existence of a nonzero-divisor of degree $(0,1)$.

Lemma 3.5. *Let $X \subseteq \mathbb{P}^1 \times \mathbb{P}^1$ be a finite set of reduced points. Then there exists a form $L \in R_{1,0}$ such that $\overline{L}$ is a nonzero-divisor in $R/I(X)$.*

Proof. Let $Y = \{A_1,\dots,A_h\}$ be the distinct first coordinates that appear in X, and view Y as a set of points in $\mathbb{P}^1$. Let L be any linear form of $k[x_0,x_1]$ that does not vanish at any of the points in Y. Because k is infinite, such a linear form exists. We now view L as an element of $R = k[x_0,x_1,y_0,y_1]$, and thus $L \in R_{1,0}$.

Suppose that $FL \in I(X)$. For any $P \in X$, we have $F(P)L(P) = 0$. Now $P = A \times B$ with $A \in Y$. Furthermore, $L(P) = L(A)$ since L is a polynomial only in the variables $\{x_0,x_1\}$. By our choice of L, $L(A) \neq 0$. So $F(P) = 0$ for all $P \in X$, whence $F \in I(X)$. So $\overline{L}$ is not a zero-divisor in $R/I(X)$.

3.2 Representing points in $\mathbb{P}^1 \times \mathbb{P}^1$ and combinatorial descriptions

We now introduce a way to "draw" sets of points in $\mathbb{P}^1 \times \mathbb{P}^1$ and describe how to use this picture to give a combinatorial description of the points. Later in this monograph, this combinatorial information is related to some of the algebraic invariants of the associated bigraded coordinate ring $R/I(X)$.

On $\mathbb{P}^1 \times \mathbb{P}^1$ there exist two families of lines $\{H_C\}$ and $\{V_C\}$, each parametrized by $C \in \mathbb{P}^1$, with the property that if $A \neq B \in \mathbb{P}^1$, then $H_A \cap H_B = \emptyset$ and $V_A \cap V_B = \emptyset$, and for all $A, B \in \mathbb{P}^1$, $H_A \cap V_B = A \times B$ is a point on $\mathbb{P}^1 \times \mathbb{P}^1$. We can thus view $\mathbb{P}^1 \times \mathbb{P}^1$ as a grid with horizontal and vertical rulings. A point $P = [a_0 : a_1] \times [b_0 : b_1] \in \mathbb{P}^1 \times \mathbb{P}^1$ can be viewed as the intersection of the horizontal ruling defined by the degree $(1,0)$ line $H = a_1x_0 - a_0x_1$ and the vertical ruling defined by the degree $(0,1)$ line $V = b_1y_0 - b_0y_1$. Hence, to any nonempty finite set $X \subset \mathbb{P}^1 \times \mathbb{P}^1$ of points we associate a set L_X of integer lattice points indicating which points lie on the same horizontal or vertical ruling. The idea is to enumerate the horizontal and vertical rulings whose intersection with X is non-empty. We thus obtain, say, $H_1, \ldots, H_h$ and $V_1, \ldots, V_v$ where $X \subset \bigcup_{i=1}^h H_i$ and $X \subset \bigcup_{j=1}^v V_j$, and L_X consists of all pairs (i,j) such that $X \cap H_i \cap V_j \neq \varnothing$.

If $\pi_1 : \mathbb{P}^1 \times \mathbb{P}^1 \to \mathbb{P}^1$ denotes the natural projection morphism onto the first coordinate, then note that $h = |\pi_1(X)|$, the number of distinct first coordinates that appear in X. Similarly, if $\pi_2 : \mathbb{P}^1 \times \mathbb{P}^1 \to \mathbb{P}^1$ is the projection morphism onto the second coordinate, then $v = |\pi_2(X)|$ is the number of distinct second coordinates. By drawing the h horizontal rulings, indexed by the first coordinates of the points appearing in X, and by drawing the v vertical rulings indexed by the second coordinates, the set X is a subset of the points of intersections of these rulings. Note that while one may normally associate the first coordinate of a point with a vertical line and the second coordinate with a horizontal line, we wish to use the reverse orientation because we wish to emulate matrix notation and the notation of a Ferrers diagram (see below). This point-of-view will simplify some of our future arguments.

We illustrate these ideas with two examples.

Example 3.6. Let $A, B \in \mathbb{P}^1$ be two distinct points, and let $X = \{A \times A, B \times B\}$. Then $|\pi_1(X)| = 2$ and $|\pi_2(X)| = 2$. So, we have two horizontal rulings, indexed by the elements of $\pi_1(X) = \{A, B\}$, i.e., H_A and H_B, and similarly, there are two vertical rulings, V_A and V_B. We can visualize X as the set of points in Figure 3.1. The dots represent the points of X. The two points are non-collinear because there is no line, that is, neither a degree $(1,0)$ form nor a degree $(0,1)$ form, that contains both points.

Example 3.7. We consider a more complicated example. If

$$X = \{A_1 \times B_2, A_1 \times B_3, A_2 \times B_1, A_2 \times B_3, A_2 \times B_4, A_3 \times B_1, A_3 \times B_4\} \subseteq \mathbb{P}^1 \times \mathbb{P}^1,$$

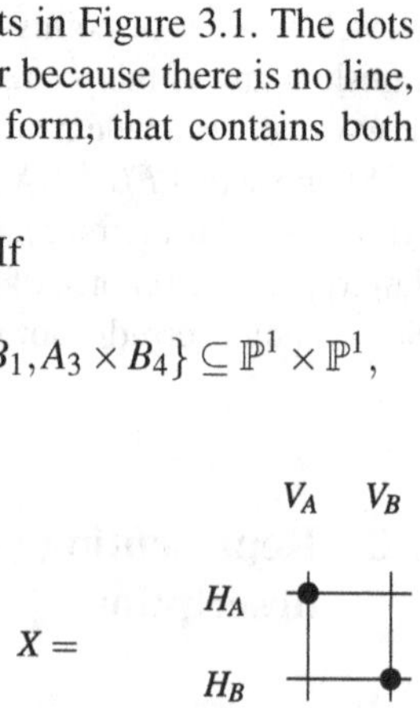

Fig. 3.1 Two non-collinear points in $\mathbb{P}^1 \times \mathbb{P}^1$

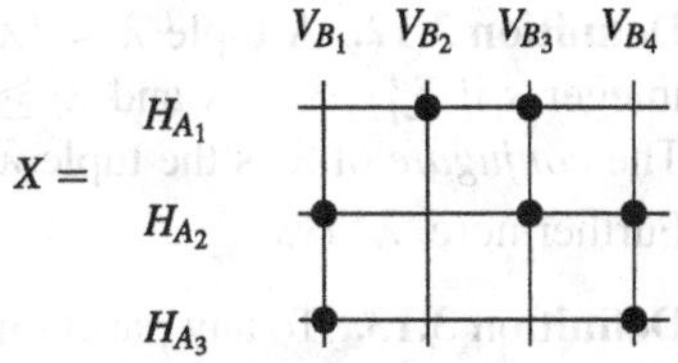

Fig. 3.2 Seven points in $\mathbb{P}^1 \times \mathbb{P}^1$

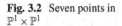

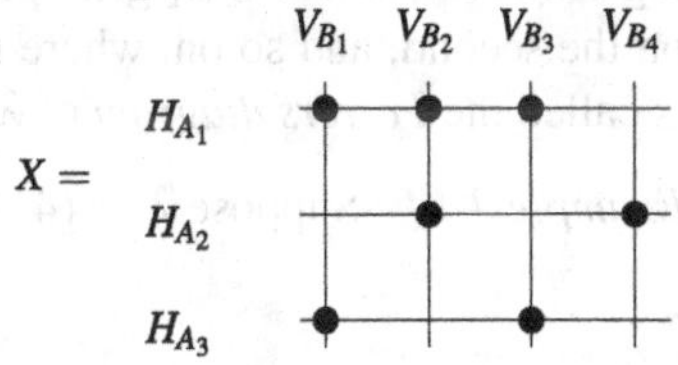

Fig. 3.3 Points of Figure 3.2 redrawn with respect to Convention 3.9

then there are $|\pi_1(X)| = 3$ horizontal rulings indexed by $\{A_1, A_2, A_3\}$ and there are $|\pi_2(X)| = 4$ vertical rulings indexed by $\{B_1, B_2, B_3, B_4\}$. So we can visualize X as in Figure 3.2, where the dots represent the points in X.

We make two conventions for moving forward.

Convention 3.8. We shall abuse notation to let H_A, respectively V_B, denote both the horizontal ruling, respectively the vertical ruling, and the degree $(1,0)$ form, respectively the degree $(0,1)$ form, that defines the ruling. Thus, given a point $P = A \times B \in \mathbb{P}^1 \times \mathbb{P}^1$, its defining ideal is given by $I(P) = (H_A, V_B)$, and geometrically, $P = H_A \cap V_B$.

Convention 3.9. By relabelling the horizontal and vertical rulings, we can always assume that $|X \cap H_{A_1}| \geq |X \cap H_{A_2}| \geq \cdots$ and $|X \cap V_{B_1}| \geq |X \cap V_{B_2}| \geq \cdots$. That is, we can assume that the first ruling contains the most number of points, the second ruling contains the same number or less, and so on.

Example 3.10. Our pictures of $X \subseteq \mathbb{P}^1 \times \mathbb{P}^1$ will be "drawn" to reflect Convention 3.9. The set of points of Example 3.7 relabelled according to Convention 3.9 can be drawn as in Figure 3.3. So, we have three points on the first horizontal ruling, two on the second, and two on the third. For the vertical rulings, we have two points on the first, second, and third vertical rulings, and one point on the fourth ruling.

With this notation and these conventions, we introduce the first of two combinatorial descriptions of a set of points in $\mathbb{P}^1 \times \mathbb{P}^1$.

Definition 3.11. Let $X \subseteq \mathbb{P}^1 \times \mathbb{P}^1$ be a finite set of reduced points and suppose that $\pi_1(X) = \{A_1, \ldots, A_h\}$ and $\pi_2(X) = \{B_1, \ldots, B_v\}$. For $i = 1, \ldots, h$, set $\alpha_i := |\pi_1^{-1}(A_i) \cap X|$, and let $\alpha_X := (\alpha_1, \ldots, \alpha_h)$. Similarly, for $j = 1, \ldots, v$, set $\beta_j := |\pi_2^{-1}(B_j) \cap X|$ and $\beta_X := (\beta_1, \ldots, \beta_v)$.

The number α_i counts the number of points in X whose first coordinate is A_i. By Convention 3.9, we have $\alpha_1 \geq \alpha_2 \geq \cdots \geq \alpha_h$, and similarly, $\beta_1 \geq \cdots \geq \beta_v$. Every set of points X can now be associated with two tuples α_X and β_X. Both α_X and β_X are partitions of $|X|$. We recall this definition, and a related concept, the conjugate of a partition.

Definition 3.12. A tuple $\lambda = (\lambda_1, \ldots, \lambda_r)$ of positive integers is a *partition* of an integer s if $\sum_{i=1}^{r} \lambda_i = s$ and $\lambda_i \geq \lambda_{i+1}$ for every i. We write $\lambda = (\lambda_1, \ldots, \lambda_r) \vdash s$. The *conjugate* of λ is the tuple $\lambda^* = (\lambda_1^*, \ldots, \lambda_{\lambda_1}^*)$ where $\lambda_i^* = \#\{\lambda_j \in \lambda \mid \lambda_j \geq i\}$. Furthermore, $\lambda^* \vdash s$.

Definition 3.13. To any partition $\lambda = (\lambda_1, \ldots, \lambda_r) \vdash s$ we can associate the following diagram: on an $r \times \lambda_1$ grid, place λ_1 points on the first horizontal line, λ_2 points on the second, and so on, where the points are left justified. The resulting diagram is called the *Ferrers diagram* of λ.

Example 3.14. Suppose $\lambda = (4,4,3,1,1) \vdash 13$. Then the Ferrers diagram of λ is

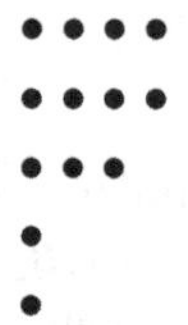

The conjugate of λ can be read off the Ferrers diagram of λ by counting the number of dots in each column as opposed to each row. In this example $\lambda^* = (5,3,3,2)$.

Definition 3.15. Let $X \subseteq \mathbb{P}^1 \times \mathbb{P}^1$ be a finite set of reduced points. We say that X *resembles a Ferrers diagram* if after we apply Convention 3.9, the set of points looks like a Ferrers diagram.

Example 3.16. Consider the set of points in Figure 3.4. After relabelling the points with respect to Convention 3.9, the set of points X resembles the Ferrers diagram of the partition $\alpha_X = (6,5,3,1,1)$. The relabelled points appear in Figure 3.5.

One can deduce the following facts directly from the definitions.

Lemma 3.17. *Let $X \subseteq \mathbb{P}^1 \times \mathbb{P}^1$ be a finite set of reduced points with associated tuples $\alpha_X = (\alpha_1, \ldots, \alpha_h)$ and $\beta_X = (\beta_1, \ldots, \beta_v)$. Then*

(*i*) $\alpha_X \vdash |X|$ *and* $\beta_X \vdash |X|$.
(*ii*) $\alpha_1^* = h$ *and* α_X^* *has length* α_1, *i.e.,* $\alpha_X^* = (h, \alpha_2^*, \ldots, \alpha_{\alpha_1}^*)$.
(*iii*) $\beta_1^* = v$ *and* β_X^* *has length* β_1, *i.e.,* $\beta_X^* = (v, \beta_2^*, \ldots, \beta_{\beta_1}^*)$.
(*iv*) *If X resembles a Ferrers diagram, then* $\alpha_X^* = \beta_X$ *and* $\beta_X^* = \alpha_X$.

Fig. 3.4 A set of points prior to applying Convention 3.9

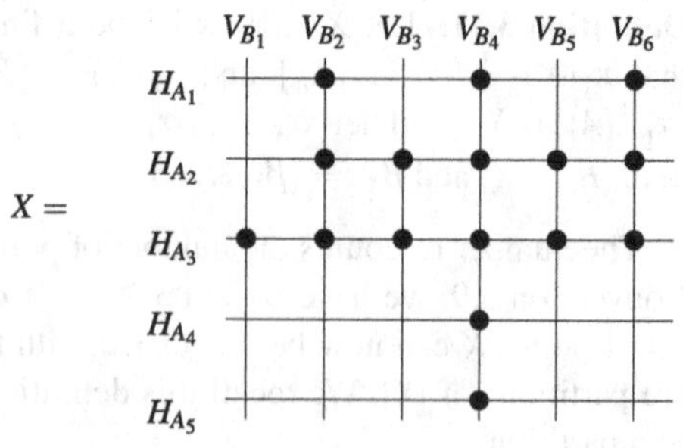

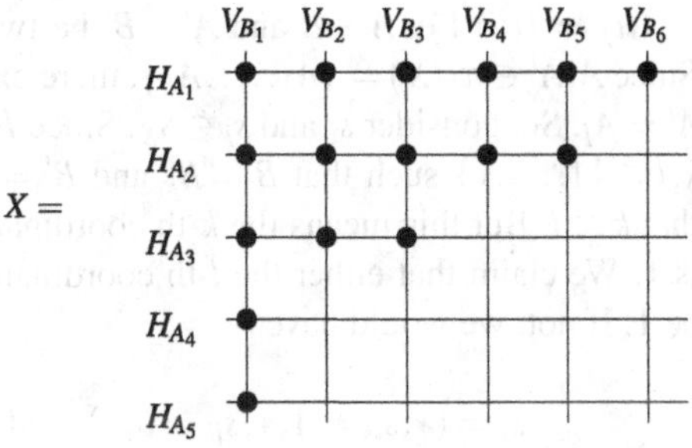

Fig. 3.5 The set of points of Figure 3.4 resembling the partition $(6,5,3,1,1)$ after relabelling

We will also exploit an alternative combinatorial description of a set of points in $\mathbb{P}^1 \times \mathbb{P}^1$.

Definition 3.18. Let $X \subseteq \mathbb{P}^1 \times \mathbb{P}^1$ be a finite set of reduced points, and let $H_{A_1}, \ldots, H_{A_h}$ and $V_{B_1}, \ldots, V_{B_v}$ be the horizontal and vertical rulings that have a non-empty intersection with X. For $i = 1, \ldots, h$, let $s_i = (s_{i,1}, \ldots, s_{i,v}) \in \mathbb{N}^v$ be a v-tuple where $s_{i,j} = 1$ if $H_{A_i} \cap V_{B_j} \in X$, and 0 otherwise. Set $S_X = \{s_1, \ldots, s_h\}$.

Because each element of S_X belongs to $\mathbb{N}^v$, we can order the elements with respect to the partial order $\succeq$ where $(u_1, \ldots, u_v) \succeq (w_1, \ldots, w_v)$ if and only if $u_i \geq v_i$ for $i = 1, \ldots, v$.

From these two combinatorial descriptions (the tuples α_X and β_X, and the set S_X), we can deduce whether or not a set of points X in $\mathbb{P}^1 \times \mathbb{P}^1$ satisfies the following geometric property.

Definition 3.19. A set of points $X \subseteq \mathbb{P}^1 \times \mathbb{P}^1$ satisfies *property* $(\star)$ if whenever $A \times B$ and $A' \times B'$ are in X with $A \neq A'$ and $B \neq B'$, then either $A \times B'$ or $A' \times B$ (or both) is also in X.

Example 3.20. The non-collinear points of Figure 3.1 fail to satisfy property $(\star)$, while the points of Figure 3.5 satisfy property $(\star)$.

The following theorem links these concepts.

Theorem 3.21. *Let $X \subseteq \mathbb{P}^1 \times \mathbb{P}^1$ be a finite set of reduced points. With the notation as above, the following are equivalent:*

(i) $\alpha_X = \beta_X^*$ *(or* $\alpha_X^* = \beta_X$*).*
(ii) *S_X has no incomparable elements with respect to the partial order $\succeq$ on $\mathbb{N}^v$.*
(iii) *X satisfies property $(\star)$.*

Proof. (i) $\Rightarrow$ (ii) If $\alpha_X = \beta_X^*$, then after we have relabelled the rulings according to Convention 3.9, the set of points will resemble a Ferrers diagram of the partition $\alpha_X = (\alpha_1, \ldots, \alpha_h)$. But this implies that for each $i = 1, \ldots, h$,

$$s_i = (\underbrace{1, \ldots, 1}_{\alpha_i}, \underbrace{0, \ldots, 0}_{(v-\alpha_i)}) \in \mathbb{N}^v.$$

Consequently $S_X = \{s_1, \ldots, s_h\}$ has no incomparable elements with respect to $\succeq$.

(ii) $\Rightarrow$ *(iii)* Let $A \times B$ and $A' \times B'$ be two points of X with $A \neq A'$ and $B \neq B'$. Since $A, A' \in \pi_1(X) = \{A_1, \ldots, A_h\}$, there exist $i, j \in \{1, \ldots, h\}$ such that $A = A_i$ and $A' = A_j$. So, consider s_i and $s_j \in S_X$. Since $B, B' \in \pi_2(X) = \{B_1, \ldots, B_\nu\}$, there exist $k, l \in \{1, \ldots, \nu\}$ such that $B = B_k$ and $B' = B_l$. Without loss of generality, assume that $k < l$. But this means the k-th coordinate of s_i is 1 and the l-th coordinate of s_j is 1. We claim that either the l-th coordinate of s_i or the k-th coordinate of s_j must be 1. If not, we would have

$$s_i = (*, s_{ik} = 1, *, s_{il} = 0, *) \text{ and } s_j = (*, s_{jk} = 0, *, s_{jl} = 1, *)$$

where $*$ denotes the other values in the ν-tuple. But then s_i and s_j are incomparable with respect to $\succeq$, contradicting our hypotheses. The presence of the 1 then implies that either $A \times B'$ or $A' \times B$ must also be an element of X, and thus X satisfies property $(\star)$.

(iii) $\Rightarrow$ *(i)* Suppose $\alpha_X = (\alpha_1, \ldots, \alpha_h)$. We proceed by induction on h. If $h = 1$, then $\alpha_X = (\alpha_1)$, which means that a single horizontal ruling contains all α_1 points of X. Now X satisfies *(iii)* vacuously, and in this case $\beta_X^* = (\underbrace{1, \ldots, 1}_{\alpha_1}) = \alpha_X$.

Assume $h > 1$. Let H_{A_h} be the horizontal ruling that contains α_h points of X, and let $X' \subseteq X$ be the set of points on the other $h - 1$ rulings. Note that X' also satisfies property $(\star)$, so by induction $\alpha_{X'} = (\alpha_1, \ldots, \alpha_{h-1}) = \beta_{X'}^*$. After relabelling, the set of points X' resembles the Ferrers diagram of $\alpha_{X'}$.

Let $A_h \times B_{i_1}, \ldots, A_h \times B_{i_{\alpha_h}}$ be the α_h points on the horizontal ruling H_{A_h}. We now claim that for each $j = 1, \ldots, h-1$, the points $A_j \times B_{i_1}, \ldots, A_j \times B_{i_{\alpha_h}}$ also belong to X'. Indeed, suppose that there is some j and i_k such that $A_j \times B_{i_k} \notin X$. Let $A_j \times B_{l_1}, \ldots, A_j \times B_{l_{\alpha_j}}$ be the α_j points on the horizontal ruling H_{A_j}. Since X satisfies property $(\star)$, and because $A_j \times B_{i_k} \notin X$, we must have $A_h \times B_{i_l}, \ldots, A_h \times B_{l_{\alpha_j}}$ all on H_{A_h}. So, H_{A_h} contains at least $\alpha_j + 1$ points, that is, $\alpha_h \geq \alpha_j + 1$. But this is a contradiction because we then have $\alpha_j \geq \alpha_h \geq \alpha_j + 1$.

We relabel the vertical rulings so that $V_{B_{i_1}}, \ldots, V_{B_{i_{\alpha_h}}}$ are the first α_h vertical rulings $V_{B_1}, \ldots, V_{B_{\alpha_h}}$. After this relabelling, X' still resembles a Ferrers diagram, and X resembles the Ferrers diagram formed by adding α_h dots to the bottom row of the Ferrers diagram of X'. So X resembles a Ferrers diagram of the partition of α_X, from which it follows that $\alpha_X = \beta_X^*$.

Remark 3.22. If X satisfies one of the equivalent conditions of Theorem 3.21, after we relabel the rulings with respect to Convention 3.9, the set of points will resemble a Ferrers diagram. On the other hand, if X fails to satisfy the conditions of Theorem 3.21, then no relabelling will enable the points to resemble a Ferrers diagram.

Example 3.23. We illustrate the above ideas by returning to the set of points X in Example 3.10. For this set of points, $\alpha_X = (3,2,2)$ and $\beta_X = (2,2,2,1)$. Because $\alpha_X^* = (3,3,1)$, we have $\alpha_X^* \neq \beta_X$.

By Theorem 3.21, the set S_X has incomparable elements and X fails property $(\star)$. Indeed, for this set of points

$$S_X = \{(1,1,1,0),(0,1,0,1),(1,0,1,0)\}.$$

The last two tuples are incomparable with respect to $\succeq$. This example also fails property $(\star)$ because $A_2 \times B_4$ and $A_3 \times B_3$ are in X, but neither $A_2 \times B_3$ nor $A_3 \times B_4$ belong to X.

Example 3.24. As observed, the set of points X of Figure 3.5 resembles the Ferrers diagram of $(6,5,3,1,1)$. The reader can check that the other equivalent conditions of Theorem 3.21 hold for this set of points.

3.3 Hilbert functions of points in $\mathbb{P}^1 \times \mathbb{P}^1$: first properties

We describe some properties of the Hilbert functions for any finite set of points in $\mathbb{P}^1 \times \mathbb{P}^1$. In particular, we show how the combinatorial information from the previous section is encoded into the bigraded Hilbert function.

We begin with a lemma about sets of points in $\mathbb{P}^1$ and a lemma about a special configuration of points in $\mathbb{P}^1 \times \mathbb{P}^1$.

Lemma 3.25. *If X is a set of s points in $\mathbb{P}^1$, then the Hilbert function of X is given by*

$$H_X(i) = \begin{cases} i+1 & i < s \\ s & i \geq s. \end{cases}$$

Proof. Let $X = \{P_1,\ldots,P_s\}$ be the s points of X. For each $j = 1,\ldots,s$, $I(P_j)$ is a principal ideal, i.e., $I(P_j) = (L_{P_j})$ for some linear form L_{P_j} in the ring $S = k[x_0,x_1]$. Then $I(X) = \bigcap_{j=1}^{s}(L_{P_j}) = (L_{P_1}\cdots L_{P_s})$.

Set $F = L_{P_1}\cdots L_{P_s}$. The result now follows from the short exact sequence

$$0 \longrightarrow S(-s) \xrightarrow{\times F} S \longrightarrow S/I(X) = S/(F) \longrightarrow 0$$

where the third map is the natural homomorphism $G \mapsto G + (F)$, and the fact that $\dim_k S_i = i+1$ for all $i \geq 0$.

We adapt the proof of Lemma 3.25 to prove a similar result for points on one ruling in $\mathbb{P}^1 \times \mathbb{P}^1$. This result shall be useful for future inductive proofs.

Lemma 3.26. *Let $X \subseteq \mathbb{P}^1 \times \mathbb{P}^1$ be a finite set of distinct points of the form*

$$X = \{A \times B_1, A \times B_2, \ldots, A \times B_v\}.$$

Then

$$H_X = \begin{bmatrix} 1 & 2 & \cdots & v-1 & v & v & \cdots \\ 1 & 2 & \cdots & v-1 & v & v & \cdots \\ 1 & 2 & \cdots & v-1 & v & v & \cdots \\ \vdots & \vdots & \vdots & \vdots & \vdots & \vdots & \ddots \end{bmatrix}.$$

Proof. After a change of coordinates, we can assume that $A = [1:0]$. So,

$$I(X) = \bigcap_{j=1}^{v} I(A \times B_j) = \bigcap_{j=1}^{v} (x_1, V_{B_j}) = (x_1, V_{B_1} \cdots V_{B_v}).$$

Thus

$$R/I(X) \cong k[x_0, y_0, y_1]/(V_{B_1} \cdots V_{B_v}).$$

Let $S = k[x_0, y_0, y_1]$ with $\deg x_0 = (1,0)$ and $\deg y_i = (0,1)$ for $i = 0,1$, and set $F = V_{B_1} \cdots V_{B_v}$. The conclusion now follows from the short exact sequence

$$0 \longrightarrow S(0,-v) \xrightarrow{\times F} S \longrightarrow S/(F) \longrightarrow 0$$

and the fact that $\dim_k S_{i,j} = j+1$ for all $(i,j) \in \mathbb{N}^2$.

Theorem 3.27. *Let $X \subseteq \mathbb{P}^1 \times \mathbb{P}^1$ be a finite set of distinct points with Hilbert function H_X.*

(i) The sequence $\{H_X(i,0)\}_{i \in \mathbb{N}}$ is the Hilbert function of $\pi_1(X) \subseteq \mathbb{P}^1$.
(ii) The sequence $\{H_X(0,j)\}_{j \in \mathbb{N}}$ is the Hilbert function of $\pi_2(X) \subseteq \mathbb{P}^1$.
(iii) $H_X(i,j) \leq H_X(i+1,j)$ for all $(i,j) \in \mathbb{N}^2$.
(iv) $H_X(i,j) \leq H_x(i,j+1)$ for all $(i,j) \in \mathbb{N}^2$.
(v) If $H_X(i,j) = H_X(i+1,j)$, then $H_X(i,j) = H_X(i+a,j)$ for all $a \in \mathbb{N}$.
(vi) If $H_X(i,j) = H_X(i,j+1)$, then $H_X(i,j) = H_X(i,j+b)$ for all $b \in \mathbb{N}$.

Proof. It suffices to prove statements (i), (iii), and (v) since the other statements are proved similarly, but use the other grading.

(i) Let $I = I(\pi_1(X)) \subseteq S = k[x_0, x_1]$. We wish to show that $(R/I(X))_{i,0} \cong (S/I)_i$ for all $i \in \mathbb{N}$. Since $R_{i,0} \cong S_i$ for all $i \in \mathbb{N}$, it is enough to show that $(I(X))_{i,0} \cong I_i$ for all $i \in \mathbb{N}$.

If $P = A \times B$ is a point of $X \subseteq \mathbb{P}^1 \times \mathbb{P}^1$, then by Theorem 3.1 the ideal associated with P has the form $I(P) = (H, V)$ with $\deg H = (1,0)$ and $\deg V = (0,1)$. Since $\pi_1(\{P\}) = \{A\}$, the ideal associated with A in S is $I(A) = (H)$, where we consider H as an $\mathbb{N}^1$-graded element of S. Then there is an isomorphism of vector spaces $I(P)_{i,0} = (H)_{i,0} \cong I(A)_i$ for each positive integer i.

Thus, if $X = \{P_1, \ldots, P_s\}$, then $\pi_1(X) = \{\pi_1(P_1), \ldots, \pi_1(P_s)\}$, and hence

$$(I(X))_{i,0} = \bigcap_{j=1}^{s} (I_{P_j})_{i,0} \cong \bigcap_{j=1}^{s} (I(\pi_1(P_j)))_i = I_i \quad \text{for all } i \in \mathbb{N}.$$

(*iii*) By Lemma 3.5 there exists a form $L \in R_{1,0}$ that is a nonzero-divisor of $R/I(X)$. Because $\deg L = (1,0)$, we have the following map between vector spaces

$$\times L : (R/I(X))_{i,j} \xrightarrow{\times \overline{L}} (R/I(X))_{i+1,j}$$

that is injective for all $(i,j) \in \mathbb{N}^2$. Because the map is injective, we must have $H_X(i,j) \leq H_X(i,j+1)$.

(*v*) Let $L \in R_{1,0}$ be the nonzero-divisor of $R/I(X)$ of Lemma 3.5. For each $(i,j) \in \mathbb{N}^2$, we have the following short exact sequence of vector spaces:

$$0 \longrightarrow (R/I(X))_{i,j} \xrightarrow{\times \overline{L}} (R/I(X))_{i+1,j} \longrightarrow (R/(I(X),L))_{i+1,j} \longrightarrow 0.$$

If $H_X(i,j) = H_X(i+1,j)$, then the multiplication map $\times \overline{L}$ is an isomorphism of vector spaces, and thus, $(R/(I(X),L))_{i+1,j} = 0$. But then for any $a \geq 1$, we also have $(R/(I(X),L))_{i+a,j} = 0$. Hence, from the short exact sequence

$$0 \longrightarrow (R/I(X))_{i+a-1,j} \xrightarrow{\times \overline{L}} (R/I(X))_{i+a,j} \longrightarrow (R/(I(X),L))_{i+a,j} \longrightarrow 0$$

we deduce that $\dim_k (R/I(X))_{i,j} = \dim_k (R/I(X))_{i+a,j}$ for all $a \geq 1$.

The Hilbert function of X captures the combinatorial information contained in α_X and β_X. Theorem 3.29 explains this correspondence. We first require a lemma. Although the lemma has been stated for a nonzero-divisor of degree $(1,0)$, a similar statement can be proved for a nonzero-divisor of degree $(0,1)$.

Lemma 3.28. *Let $X \subseteq \mathbb{P}^1 \times \mathbb{P}^1$ be a finite set of distinct points with $|\pi_1(X)| = h$. Suppose that $L \in R_{1,0}$ is a nonzero-divisor on $R/I(X)$. Then $(x_0,x_1)^h \subseteq (I(X),L)$.*

Proof. We have a short exact sequence

$$0 \longrightarrow (R/I(X))(-1,0) \xrightarrow{\times \overline{L}} R/I(X) \longrightarrow R/(I(X),L) \longrightarrow 0$$

because L is a nonzero-divisor, from which we deduce that

$$H_{R/(I(X),L)}(i,j) = H_X(i,j) - H_X(i-1,j) \text{ for all } (i,j) \in \mathbb{N}^2.$$

By Theorem 3.27 (*i*), the sequence $\{H_X(i,0)\}_{i \in \mathbb{N}}$ is the Hilbert function of $\pi_1(X) \subseteq \mathbb{P}^1$. Because $|\pi_1(X)| = h$, we have $H_X(i,0) = h$ for all $i \geq h-1$ by Lemma 3.25. Thus, $H_{R/(I(X),L)}(h,0) = h - h = 0$, which means that $(I(X),L)_{h,0} = R_{h,0}$. This is equivalent to the statement that $(x_0,x_1)^h \subseteq (I(X),L)$.

Theorem 3.29. *Let $X \subseteq \mathbb{P}^1 \times \mathbb{P}^1$ be any set of distinct points with associated tuples $\alpha_X = (\alpha_1,\ldots,\alpha_h)$ and $\beta_X = (\beta_1,\ldots,\beta_v)$, and let $h = |\pi_1(X)|$ and $v = |\pi_2(X)|$.*

(*i*) *For all* $j \in \mathbb{N}$, *if* $i \geq h-1$, *then*

$$H_X(i,j) = \alpha_1^* + \alpha_2^* + \cdots + \alpha_{j+1}^*$$

where $\alpha_X^* = (\alpha_1^*, \ldots, \alpha_{\alpha_1}^*)$ *is the conjugate of* α_X, *and where we make the convention that* $\alpha_l^* = 0$ *if* $l > \alpha_1$.

(*ii*) *For all* $i \in \mathbb{N}$, *if* $j \geq v-1$, *then*

$$H_X(i,j) = \beta_1^* + \beta_2^* + \cdots + \beta_{i+1}^*$$

where $\beta_X^* = (\beta_1^*, \ldots, \beta_{\beta_1}^*)$ *is the conjugate of* β_X, *and where we make the convention that* $\beta_l^* = 0$ *if* $l > \beta_1$.

Proof. We only prove (*i*) because the proof of the second statement is the same, but takes into account the other grading.

We prove the statement by induction on h. Suppose that $h = 1$. In this case, $X = \{A \times B_1, \ldots, A \times B_v\}$, and thus $\alpha_X = (v)$, and $\alpha_X^* = \underbrace{(1, \ldots, 1)}_{v}$. By Lemma 3.26, the Hilbert function of X is given by

$$H_X = \begin{bmatrix} 1 & 2 & \cdots & v-1 & v & v & \cdots \\ 1 & 2 & \cdots & v-1 & v & v & \cdots \\ 1 & 2 & \cdots & v-1 & v & v & \cdots \\ \vdots & \vdots & & \vdots & \vdots & \vdots & \ddots \end{bmatrix}.$$

From H_X, we see that for all $i \geq h-1 = 0$,

$$H_X(i,j) = \alpha_1^* + \alpha_2^* + \cdots + \alpha_{j+1}^* = \min\{j+1, v\} \quad \text{for all } j \in \mathbb{N}$$

since $\alpha_X^* = \underbrace{(1, \ldots, 1)}_{v}$.

Now assume that $h > 1$. If $H_1, \ldots, H_h$ are the h horizontal rulings that contain X, let X_1 be all the points of X that are contained on $H_1, \ldots, H_{h-1}$, and let X_2 be all the points of X that are contained on the horizontal ruling H_h. Thus $X = X_1 \cup X_2$.

We have the following short exact sequence

$$0 \longrightarrow R/(I(X_1) \cap I(X_2)) \longrightarrow R/I(X_1) \oplus R/I(X_2) \longrightarrow R/(I(X_1) + I(X_2)) \longrightarrow 0.$$

Note that $I(X_2) = (H_h, G)$ where G is a polynomial only in the y_is.

Since H_h does not pass through any of the points of X_1, H_h is a nonzero-divisor of degree $(1,0)$ on $R/I(X_1)$. Thus by Lemma 3.28,

$$R_{i,j} = (I(X_1), H_h)_{i,j} \subseteq (I(X_1) + I(X_2))_{i,j} \text{ if } i \geq |\pi_1(X_1)| = h-1.$$

Thus, the short exact sequence implies that

$$H_X(i,j) = H_{X_1}(i,j) + H_{X_2}(i,j) \text{ for all } (i,j) \in \mathbb{N}^2 \text{ with } i \geq h-1.$$

To complete the proof, note that if $\alpha_X = (\alpha_1, \ldots, \alpha_h)$, then $\alpha' = \alpha_{X_1} = (\alpha_1, \ldots, \alpha_{h-1})$ and $\alpha'' = \alpha_{X_2} = (\alpha_h)$. By using induction with the above formula for $H_X(i,j)$, we have

$$\begin{aligned} H_X(i,j) &= ((\alpha')_1^* + \cdots + (\alpha')_{j+1}^*) + ((\alpha'')_1^* + \cdots + (\alpha'')_{j+1}^*) \\ &= ((\alpha')_1^* + (\alpha'')_1^*) + \cdots + ((\alpha')_{j+1}^* + (\alpha'')_{j+1}^*) \\ &= \alpha_1^* + \cdots + \alpha_{j+1}^*. \end{aligned}$$

for all $(i,j) \in \mathbb{N}^2$ with $i \geq h-1$.

By combining Theorem 3.29 with Theorem 3.27, all but a finite number of the values of H_X can be determined directly from the tuples α_X and β_X.

Corollary 3.30. *Let $X \subseteq \mathbb{P}^1 \times \mathbb{P}^1$ be a set of s distinct points with $\alpha_X = (\alpha_1, \ldots, \alpha_h)$ and $\beta_X = (\beta_1, \ldots, \beta_v)$. For each $i \geq 1$ and $j \geq 1$, set*

$$a_i = \sum_{l=1}^{i} \alpha_l^* \text{ and } b_j = \sum_{l=1}^{j} \beta_l^*$$

where $\alpha_X^* = (\alpha_1^*, \ldots, \alpha_{\alpha_1}^*)$ and $\beta_X^* = (\beta_1^*, \ldots, \beta_{\beta_1}^*)$, and we adopt the convention that $\alpha_l^* = 0$ if $l > \alpha_1$ and $\beta_k^* = 0$ if $k > \beta_1$. Then

$$H_X = \begin{bmatrix} 1 & 2 & 3 & \cdots & v-1 & b_1 & b_1 & \cdots \\ 2 & & & & & b_2 & b_2 & \cdots \\ 3 & & & & & b_3 & b_3 & \cdots \\ \vdots & & & & & \vdots & & \\ h-1 & & & & & b_{h-1} & b_{h-1} & \cdots \\ a_1 & a_2 & a_3 & \cdots & a_{v-1} & s & s & \cdots \\ a_1 & a_2 & a_3 & \cdots & a_{v-1} & s & s & \cdots \\ \vdots & \vdots & \vdots & \cdots & \vdots & \vdots & \vdots & \ddots \end{bmatrix}.$$

Proof. The entries in the top row and the left most column follow from Theorem 3.27 (i) and (ii) because the left most column is the Hilbert function of h points in $\mathbb{P}^1$ and the top row is the Hilbert function of v points in $\mathbb{P}^1$. Note that by Lemma 3.17, we have $h = a_1$ and $v = b_1$.

The entries in columns $j = 2, \ldots, v-1$ are given by Theorem 3.29. The entries in rows $i = 2, \ldots, h-1$ are also a consequence of this theorem. Finally, note that for all $(i,j) \geq (h-1, v-1)$, $\alpha_1^* + \cdots + \alpha_{\alpha_1}^* = \beta_1^* + \cdots + \beta_{\beta_1}^* = s$.

Remark 3.31. It follows from Corollary 3.30 that all the values of the Hilbert function of X can be computed from α_X and β_X except for the values $H_X(i,j)$ with $(1,1) \preceq (i,j) \preceq (h-2, v-2)$. Thus, if either $h \leq 2$ or $v \leq 2$, then the entire Hilbert function can be computed from α_X and β_X.

Example 3.32. For the points of Example 3.7 we have $\alpha_X = (3,2,2)$ and $\beta_X = (2,2,2,1)$. For these points $\alpha_X^* = (3,3,1)$ and $\beta_X^* = (4,3)$. So $a_1 = 3, a_2 = 6, a_3 = 7 = a_4 = a_5 = \cdots$ and $b_1 = 4$, $b_2 = 7 = b_3 = b_4 = \cdots$. Hence, for this set of points, we know the following values of H_X:

$$H_X = \begin{bmatrix} 1 & 2 & 3 & 4 & 4 & \cdots \\ 2 & ? & ? & 7 & 7 & \cdots \\ 3 & 6 & 7 & 7 & 7 & \cdots \\ 3 & 6 & 7 & 7 & 7 & \cdots \\ \vdots & \vdots & \vdots & \vdots & \vdots & \ddots \end{bmatrix}.$$

The values of H_X denoted by ?, namely, $H_X(1,1)$ and $H_X(1,2)$, cannot be determined from α_X and β_X.

Example 3.33. It is possible for two sets of points to have the same α_X and β_X, but not the same Hilbert function. For example, let A_1, A_2, A_3 be three distinct points of $\mathbb{P}^1$, and let B_1, B_2, and B_3 be another collection of three distinct points in $\mathbb{P}^1$. Let $X_1 = \{A_1 \times B_1, A_2 \times B_2, A_2 \times B_3, A_3 \times B_1\}$, and let $X_2 = \{A_1 \times B_3, A_2 \times B_1, A_2 \times B_2, A_3 \times B_1\}$. We can visualize these two sets as in Figure 3.6.

For these two sets of points, $\alpha_{X_1} = \alpha_{X_2} = (2,1,1)$ and $\beta_{X_1} = \beta_{X_2} = (2,1,1)$. Hence, for all $(i,j) \neq (1,1)$, Corollary 3.30 implies that $H_{X_1}(i,j) = H_{X_2}(i,j)$. Using CoCoA to compute the Hilbert function of X_1 and X_2, we find that the Hilbert functions are not equal. Specifically,

$$H_{X_1} = \begin{bmatrix} 1 & 2 & 3 & \cdots \\ 2 & 3 & 4 & \cdots \\ 3 & 4 & 4 & \cdots \\ \vdots & \vdots & \vdots & \ddots \end{bmatrix} \qquad H_{X_2} = \begin{bmatrix} 1 & 2 & 3 & \cdots \\ 2 & 4 & 4 & \cdots \\ 3 & 4 & 4 & \cdots \\ \vdots & \vdots & \vdots & \ddots \end{bmatrix}.$$

Observe that the form $H_{A_2}V_{B_1}$ is a form of degree $(1,1)$ that passes through the points of X_1, thus explaining why $H_{X_1}(1,1) < 4$.

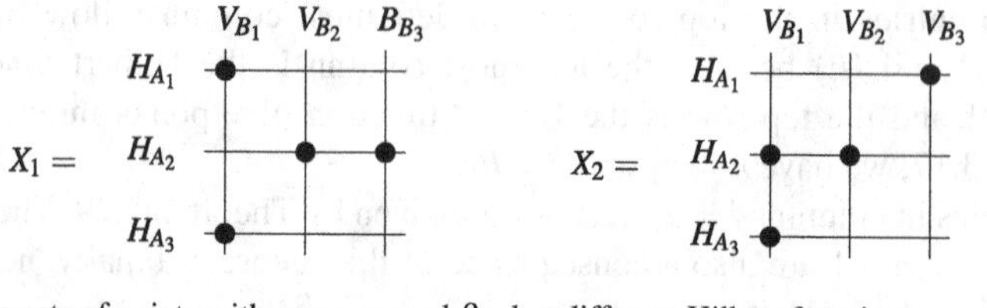

Fig. 3.6 Two sets of points with same α_X and β_X, but different Hilbert functions

Note that while we can find X_1 and X_2 that have $\alpha_{X_1} = \alpha_{X_2}$ and $\beta_{X_1} = \beta_{X_2}$, we will always have $S_{X_1} \neq S_{X_2}$ except when $X_1 = X_2$.

3.4 Separators

Given a set of points X in $\mathbb{P}^1 \times \mathbb{P}^1$, and any point $P \in X$, we sometimes want to compare the properties of X with those of $X \setminus \{P\}$. A separator gives us a tool to compare and contrast these two sets of points.

Definition 3.34. Let X be a set of distinct points in $\mathbb{P}^1 \times \mathbb{P}^1$ with $P \in X$. A bihomogeneous form $F \in R$ is a *separator for P* if $F(P) \neq 0$, but $F(Q) = 0$ for all $Q \in X \setminus \{P\}$. We call F a *minimal separator* for P if there does not exist a separator G for P with $\deg G \prec \deg F$.

If $T \subseteq \mathbb{N}^2$ is a subset, then let $\min T$ denote the set of minimal elements of T with respect to the partial order $\succeq$ on $\mathbb{N}^2$.

Definition 3.35. Let X be a set of distinct points in $\mathbb{P}^1 \times \mathbb{P}^1$. The *degree of a point* $P \in X$ is the set $\deg_X(P) = \min\{\deg F \mid F \text{ is a separator for } P \in X\}$.

Example 3.36. Consider the points of Example 3.6, i.e., let $A, B \in \mathbb{P}^1$ be two distinct points, and let $X = \{A \times A, B \times B\}$. We can visualize X as in Figure 3.7. Consider the point $P = B \times B$. For this point, the degree $(1,0)$ form H_A and the degree $(0,1)$ form V_A are both separators of P. Indeed, they pass through all the points of X except P. Because there is no constant term that vanishes at P, we have $\deg_X(P) = \{(1,0),(0,1)\}$. This gives us an example where $|\deg_X(P)| > 1$.

We can compute $\deg_X(P)$ by comparing the Hilbert functions of X and $X \setminus \{P\}$ as shown below. We require the following notation. For any $(i,j) \in \mathbb{N}^2$, define

$$D_{(i,j)} := \{(k,l) \in \mathbb{N}^2 \mid (k,l) \succeq (i,j)\}.$$

For any finite set $T \subseteq \mathbb{N}^2$, we set

$$D_T := \bigcup_{(i,j)\in T} D_{(i,j)}.$$

Note that $\min D_T = T$. Thus D_T can be viewed as the largest subset of $\mathbb{N}^2$ whose set of minimal elements is the set T.

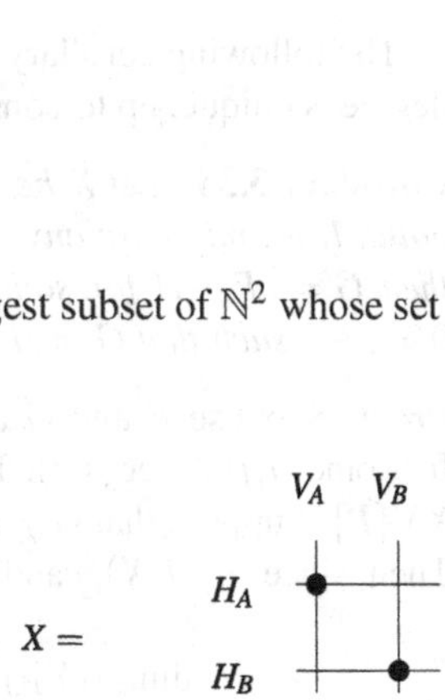

Fig. 3.7 Two non-collinear points

Theorem 3.37. *Let X be a set of distinct points in $\mathbb{P}^1 \times \mathbb{P}^1$, and let $P \in X$ be any point. If $Y = X \setminus \{P\}$, then there exists a finite set $T \subseteq \mathbb{N}^2$ such that*

$$H_Y(i,j) = \begin{cases} H_X(i,j) & \text{if } (i,j) \notin D_T \\ H_X(i,j) - 1 & \text{if } (i,j) \in D_T. \end{cases}$$

Furthermore, $\deg_X(P) = T$.

Proof. The short exact sequence

$$0 \to R/(I(Y) \cap I(P)) \to R/I(Y) \oplus R/I(P) \to R/(I(Y) + I(P)) \to 0$$

implies that

$$H_Y(i,j) = H_X(i,j) - H_P(i,j) + H_{R/(I(Y)+I(P))}(i,j) \text{ for all } (i,j) \in \mathbb{N}^2$$

because $I(Y) \cap I(P) = I(X)$.

Now $R/I(P) \cong k[x_1, y_1]$, the $\mathbb{N}^2$-graded ring with $\deg x_1 = (1,0)$ and $\deg y_1 = (0,1)$. So $H_P(i,j) = 1$ for all $(i,j) \in \mathbb{N}^2$. Also,

$$R/(I(Y) + I(P)) \cong \frac{R/I(P)}{(I(Y) + I(P))/I(P)}.$$

So, $(I(Y) + I(P))/I(P) \cong J$, where J is an $\mathbb{N}^2$-homogeneous ideal of $k[x_1, y_1]$. Thus $H_{R/(I(Y)+I(P))}(i,j) = 0$ or 1 for all $(i,j) \in \mathbb{N}^2$.

When $H_{R/(I(Y)+I(P))}(i,j) = 0$, then $H_{R/(I(Y)+I(P))}(k,l) = 0$ for all $(k,l) \succeq (i,j)$. The desired set is then $T = \min \mathscr{T}$ where $\mathscr{T} = \{(i,j) \in \mathbb{N}^2 \mid H_{R/(I(Y)+I(P))}(i,j) = 0\}$.

We now show that $\deg_X(P) = T$. Suppose that $(i,j) \in \deg_X(P)$. Thus there exists $F \in I(Y)_{i,j} \setminus I(X)_{i,j}$. So, $H_Y(i,j) < H_X(i,j)$, which means $(i,j) \in D_T$. So, there exists $(k,l) \in T$ such that $(i,j) \succeq (k,l)$. If $(i,j) \succ (k,l)$, we have $H_Y(k,l) < H_X(k,l)$, which means there is some $G \in I(Y)_{k,l}$ that is not in $I(X)_{k,l}$. Hence, G is a separator of P with a degree smaller than (i,j), contradicting our definition of $\deg_X(P)$. So $(i,j) \in T$. The reverse inclusion is proved similarly.

The following corollary shows that a minimal separator of a point P of a specific degree is unique, up to constant, in the ring $R/I(X)$.

Corollary 3.38. *Let X be a set of distinct points in $\mathbb{P}^1 \times \mathbb{P}^1$, and let $P \in X$ be any point. If F and G are any two minimal separators of P with $\deg F = \deg G = (i,j)$, then $G = cF + H$ for some $0 \neq c \in k$ and $H \in I(X)_{i,j}$. Equivalently, there exists $0 \neq c \in k$ such that $\overline{G} = c\overline{F} \in R/I(X)$.*

Proof. Suppose F and G are minimal separators of P and $\deg F = \deg G = (i,j)$ for some $(i,j) \in \deg_X(P)$. Then the vector space $(I(X), F, G)_{i,j} \subseteq I(Y)_{i,j}$ where $Y = X \setminus \{P\}$. Suppose that $G \neq cF + H$ for all nonzero scalars $c \in k$ and any $H \in I(X)_{i,j}$. Then, since $F \notin I(X)_{i,j}$ and our hypothesis on G, we must have

$$\dim_k I(Y)_{i,j} \geq \dim_k (I(X), F, G)_{i,j} \geq \dim_k I(X)_{i,j} + 2.$$

However, this inequality contradicts Theorem 3.37 which implies $\dim_k I(Y)_{i,j} \leq \dim_k I(X)_{i,j} + 1$ for all $(i,j) \in \mathbb{N}^2$. So, there exists some nonzero $c \in k$ and $H \in I(X)$ such that $G = cF + H$.

Finally, we require a result on how the dimension of a bigraded piece of the ideal $I(X)$ changes if we add a single separator to $I(X)$ as a generator.

Theorem 3.39. *Let X be a set of distinct points in $\mathbb{P}^1 \times \mathbb{P}^1$, and suppose that F is a separator of a point $P \in X$. Then*

$$\dim_k (I(X), F)_{i,j} = \dim_k I(X)_{i,j} + 1 \quad \textit{for all } (i,j) \succeq \deg F.$$

Proof. Let F be a separator of P. We first claim that $(I(X) : (F)) = I(P)$. Indeed, for any $G \in I(P)$, $FG \in I(X)$ since FG vanishes at all points of X. Conversely, let $G \in (I(X) : (F))$. So $GF \in I(X) \subseteq I(P)$. Now $F \notin I(P)$, and because $I(P)$ is a prime ideal, we have $G \in I(P)$, as desired.

To prove the statement, we consider the short exact sequence

$$0 \to R/(I(X) : (F))(-\deg F) \xrightarrow{\times F} R/I(X) \longrightarrow R/(I(X), F) \to 0.$$

Because $R/(I(X) : (F)) \cong R/I(P)$, we have

$$H_{R/(I(X),F)}(i,j) = H_X(i,j) - H_P((i,j) - \deg F) \quad \text{for all } (i,j) \in \mathbb{N}^2.$$

Now $H_P(i,j) = 1$ for all $(i,j) \in \mathbb{N}^2$, and equals 0 otherwise. The conclusion follows.

3.5 Case study

We present a case study devoted to the set of points which resemble the Ferrers diagram of $(v, 1, \ldots, 1)$. We compute the Hilbert function of this configuration, and determine the degrees of all the points in this configuration. Besides illustrating the ideas of the previous sections, these results are required in later chapters.

We construct the following family of points. Pick two integers $h \geq 1$ and $v \geq 1$. Let $\{A_1, \ldots, A_h\}$ be h distinct points in $\mathbb{P}^1$ and let $\{B_1, \ldots, B_v\}$ be v distinct points in $\mathbb{P}^1$. Construct the following $v + h - 1$ points

$$X_{h,v} := \{A_1 \times B_1, A_1 \times B_2, \ldots, A_1 \times B_v, A_2 \times B_1, A_3 \times B_1, \ldots, A_h \times B_1\}.$$

If we draw X with respect to Convention 3.9, then X resembles the Ferrers diagram of $(v, \underbrace{1, \ldots, 1}_{h-1})$, as in Figure 3.8. The Hilbert function of $X_{h,v}$ can be determined directly from v and h.

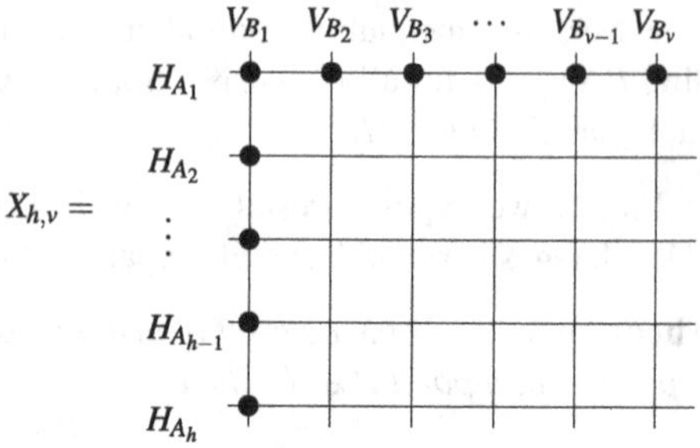

Fig. 3.8 Set of points $X_{h,v}$ that resembles the Ferrers diagram of $(v,1,\ldots,1)$

Theorem 3.40. *Fix $h \geq 1$ and $v \geq 1$, and let $X_{h,v}$ be as above. Then the Hilbert function of $X_{h,v}$ is given by*

$$H_{X_{h,v}} = \begin{bmatrix} 1 & 2 & 3 & 4 & \cdots & v-1 & v & v & \cdots \\ 2 & 3 & 4 & 5 & \cdots & v & v+1 & v+1 & \cdots \\ 3 & 4 & 5 & 6 & \cdots & v+1 & v+2 & v+2 & \cdots \\ 4 & 5 & 6 & 7 & \cdots & v+2 & v+3 & v+3 & \cdots \\ \vdots & \vdots & \vdots & \vdots & & & \vdots & \vdots & \\ h-1 & h & h+1 & h+2 & \cdots & h+v-3 & h+v-2 & h+v-2 & \cdots \\ h & h+1 & h+2 & h+3 & \cdots & h+v-2 & h+v-1 & h+v-1 & \cdots \\ h & h+1 & h+2 & h+3 & \cdots & h+v-2 & h+v-1 & h+v-1 & \cdots \\ \vdots & \vdots & \vdots & \vdots & & \vdots & \vdots & \vdots & \ddots \end{bmatrix}.$$

Proof. Let $X = X_{h,v}$. From the construction of X, we have $\alpha_X = (v, \underbrace{1,\ldots,1}_{h-1})$ and $\beta_X = (h, \underbrace{1,\ldots,1}_{v-1})$. By Remark 3.31, if $h \leq 2$ or $v \leq 2$, then we can compute H_X directly from α_X and β_X using Corollary 3.30. In these cases, H_X agrees with the statement. So, we can assume that $h > 2$ and $v > 2$.

We can use Corollary 3.30 to compute $H_X(i,j)$ when either $i \geq h-1$ or $j \geq v-1$. The values of $H_X(i,0)$ for all $i \geq 0$ and $H_X(0,j)$ for all $j \geq 0$ also follow from this corollary. It thus suffices to compute $H_X(i,j)$ for all (i,j) that satisfy $(1,1) \preceq (i,j) \preceq (h-2, v-2)$.

We claim that

$$I(X)_{i,j} = (H_{A_1} V_{B_1})_{i,j} \text{ for all } (1,1) \preceq (i,j) \preceq (h-2, v-2).$$

Because $H_{A_1} V_{B_1}$ is a degree $(1,1)$ form that belongs to $I(X)_{1,1}$, it suffices to verify that $I(X)_{i,j} \subseteq (H_{A_1} V_{B_1})_{i,j}$ for (i,j) in this range. So, suppose that $F \in I(X)_{i,j}$. Then F vanishes at the v points on the ruling H_{A_1} with $j < v-1$. By Theorem 2.15, we have $F = F' H_{A_1}$. Now F' must vanish at the $h-1$ points on the ruling V_{B_1}. Because $i < h-1$, Theorem 2.15 implies that $F' = F'' V_{B_1}$. Thus $F = F'' H_{A_1} V_{B_1} \in (H_{A_1} V_{B_1})_{i,j}$.

By this claim, it follows that for all $(1,1) \preceq (i,j) \preceq (h-2,v-2)$

$$H_X(i,j) = \dim_k R_{i,j} - \dim_k (H_{A_1} V_{B_1})_{i,j} = (i+1)(j+1) - ij = i+j+1.$$

This formula gives the desired result.

Theorem 3.41. *Fix $h \geq 1$ and $v \geq 1$, and let $X_{h,v}$ be as above. Then*

$$\deg_{X_{h,v}}(A_i \times B_j) = \begin{cases} \{(h-1,v-1)\} & \text{if } (i,j) = (1,1) \\ \{(0,v-1)\} & \text{if } (i,j) = (1,j) \text{ with } 1 < j \leq v \\ \{(h-1,0)\} & \text{if } (i,j) = (i,1) \text{ with } 1 < i \leq h. \end{cases}$$

Proof. Let $X = X_{h,v}$. We first consider the case $(i,j) = (1,1)$. By Theorem 3.37, it suffices to show that the Hilbert function of $Y = X \setminus \{A_1 \times B_1\}$ is

$$H_Y = \begin{bmatrix} 1 & 2 & 3 & 4 & \cdots & v-1 & v & v & \cdots \\ 2 & 3 & 4 & 5 & \cdots & v & v+1 & v+1 & \cdots \\ 3 & 4 & 5 & 6 & \cdots & v+1 & v+2 & v+2 & \cdots \\ 4 & 5 & 6 & 7 & \cdots & v+2 & v+3 & v+3 & \cdots \\ \vdots & \vdots & \vdots & \vdots & & \vdots & \vdots & \vdots & \\ h-1 & h & h+1 & h+2 & \cdots & h+v-3 & h+v-2 & h+v-2 & \cdots \\ h & h+1 & h+2 & h+3 & \cdots & h+v-2 & h+v-2 & h+v-2 & \cdots \\ h & h+1 & h+2 & h+3 & \cdots & h+v-2 & h+v-2 & h+v-2 & \cdots \\ \vdots & \vdots & \vdots & \vdots & & \vdots & \vdots & \vdots & \ddots \end{bmatrix}.$$

That is, we want to show that $\min\{(i,j) \mid H_X(i,j) \neq H_Y(i,j)\} = \{(h-1,v-1)\}$.

The computation of H_Y follows as in the proof of the previous theorem. We first note that $\alpha_Y = (v-1,\underbrace{1,\ldots,1}_{h-1})$ and $\beta_Y = (h-1,\underbrace{1,\ldots,1}_{v-1})$. If $h \leq 2$ or $v \leq 2$, the computation of the Hilbert function follows directly from these tuples. So, we can assume that $h > 2$ and $v > 2$. Corollary 3.30 gives us the values of $H_Y(i,j)$ if $i = 0$, or if $j = 0$, or if $i \geq h-1$, or if $j \geq v-1$. It suffices to evaluate $H_Y(i,j)$ for all $(1,1) \preceq (i,j) \preceq (h-2,v-2)$.

Even though the point $A_1 \times B_1$ has been removed from X, the form $H_{A_1} V_{B_1}$ is still in $I(Y)_{1,1}$. As in the previous proof, we claim that $I(Y)_{i,j} = (H_{A_1} V_{B_1})_{i,j}$ in the range $(1,1) \preceq (i,j) \preceq (h-2,v-2)$. Indeed, let $F \in I(Y)_{i,j}$ with (i,j) in this range. Then F must still vanish at the $v-1$ points on the ruling H_{A_1}, and since $j < v-1$, we have $F = F'H_{A_1}$ by Theorem 2.15. Furthermore F' will vanish at the $h-1$ points on V_{B_1}, and so $F' = F''V_{B_1}$, again by Theorem 2.15. The proof now follows as in the previous proof.

We next consider the case $(i,j) = (1,j)$ with $1 < j \leq v$. We observe that renumbering the vertical lines V_{B_j} for $2 \leq j \leq v$, we have that the sets $Y_j = X_{h,v} \setminus \{A_1 \times B_j\}$ are equal to $Y_v = X_{h,v} \setminus \{A_1 \times B_v\} = X_{h,v-1}$. By applying Theorem 3.40 we have that $H_{Y_j}(0,v) = H_{Y_v}(0,v) = H_{X_{h,v}}(0,v) - 1$. We now apply Theorem 3.37 to deduce that $\deg_{X_{h,v}}(A_1 \times B_j) = \deg_{X_{h,v}}(A_1 \times B_v) = \{(0,v-1)\}$ for $2 \leq j \leq v$. A similar argument is used the for case $(i,j) = (i,1)$ with $1 < i \leq h$.

3.6 Additional notes

Giuffrida, Maggioni, and Ragusa [36–38] were among the first to study sets of points in $\mathbb{P}^1 \times \mathbb{P}^1$ and the bigraded structure of the associated coordinate ring. Prior to this work, a number of papers studied collections of points on the quadric surface in $\mathbb{P}^3$ (see, for example, the papers of Paxia, Raciti, and Ragusa [78], Raciti [82], Ragusa and Zappalà [83], and Zappalà [99]). Recall that the quadric surface is isomorphic to $\mathbb{P}^1 \times \mathbb{P}^1$ (see, for example, [60, Exercise I.2.15]). While in this monograph we are considering $\mathbb{P}^1 \times \mathbb{P}^1$ to be the ambient space, the above papers take the perspective that the points belong to $\mathbb{P}^3$, but have the additional condition that the defining ideal of the set of points contains a quadric form. As a consequence, the associated coordinate ring is no longer bigraded, but it does have the regular standard grading. Many of the above cited papers contain results about the Hilbert functions of points on these quadric surfaces.

The results of Section 3.1 can be extended naturally to sets of points in $\mathbb{P}^{n_1} \times \cdots \times \mathbb{P}^{n_r}$; see, for example, a paper of the second author [93].

In Section 3.2, we introduced two different ways one can associate combinatorial data with a set of points in $\mathbb{P}^1 \times \mathbb{P}^1$. The tuples α_X and β_X first appeared in the second author's PhD thesis [92]. We first learnt about partitions and Ferrers diagrams from Ryser's classical book [85]. The set S_X first appeared in the first author's PhD thesis [42, 43]. It should be noted that S_X was introduced in the context of fat points (to be discussed in Chapter 6). We have presented S_X in the context of reduced points. The connection to property $(\star)$ was first made explicit in [48].

Theorem 3.27 of Section 3.3 extends naturally to any set of points in $\mathbb{P}^{n_1} \times \cdots \times \mathbb{P}^{n_r}$. Again, see [93] for more details. In addition, see the preprint of Şahin and Soprunov [86] for similar results for the multigraded Hilbert function of a zero-dimensional scheme on a toric variety. Note that it follows from Theorem 3.29 that if one is first given the Hilbert function H_X of a set of points $X \subseteq \mathbb{P}^1 \times \mathbb{P}^1$, then one can determine α_X (and β_X), and consequently, some combinatorial data about X. This fact was first observed by Giuffrida, Maggioni, and Ragusa [36], and has recently been extended to $\mathbb{P}^1 \times \cdots \times \mathbb{P}^1$ by the authors [52]. The connection between the tuples α_X and β_X and H_X was first made explicit by the second author in [93]. The proof given in this monograph, however, is new. As an additional comment, even if $\alpha_X^* \neq \beta_X$, there is still a relationship between these two tuples. In the language of partitions, the tuple α_X^* will *majorize* the tuple β_X. Again, see [93] for more details.

Separators first appeared in Orecchia's work [77] on the conductor of a set of points. However, the term separator does not appear until the paper of Geramita, Kreuzer, and Robbiano [33]. Separators of points in $\mathbb{P}^1 \times \mathbb{P}^1$ were first used in the PhD thesis of the first author (see [42, 43]), although they were not explicitly defined. Properties of Hilbert functions and separators in $\mathbb{P}^1 \times \mathbb{P}^1$ can be found in the work of Marino [70] and Bonacini-Marino [9, 10]. Additional properties of separators of points in $\mathbb{P}^1 \times \mathbb{P}^1$ (and more generally, in $\mathbb{P}^{n_1} \times \cdots \times \mathbb{P}^{n_r}$) were later developed by the authors in [47, 48]. One version of Theorem 3.37 first appeared in [47]; in particular, the existence of the set T is shown. The proof that T equals $\deg_X(P)$ later appeared in [48]. Corollary 3.38 first appeared in [47]. These two results can be extended to the more general case of points in $\mathbb{P}^{n_1} \times \cdots \times \mathbb{P}^{n_r}$. The case study Theorem 3.41 can be found in [47]. Note that in [47] we need to know that this set of points is an arithmetically Cohen-Macaulay set of points; our proof avoids this hypothesis.

In the next chapter, we turn our focus to a special class of points, the so-called arithmetically Cohen-Macaulay sets of points in $\mathbb{P}^1 \times \mathbb{P}^1$. Before embarking on this path, we wish to point out there is another path which is not explored in this monograph. We say that a set of points $X \subseteq \mathbb{P}^1 \times \mathbb{P}^1$ is in *generic position* if

$$H_X(i,j) = \min\{\dim_k R_{i,j}, |X|\} \quad \text{for all } (i,j) \in \mathbb{N}^2.$$

This definition generalizes the well-known definition for points in $\mathbb{P}^n$ (see, for example, [31, Definition 2.4]), and of course, generalizes to sets of points in $\mathbb{P}^{n_1} \times \cdots \times \mathbb{P}^{n_r}$. Note that the use of "generic" here is not saying that the set of points are generic, and generic position is also different from being in general position. Instead, generic position is used to describe a property of the corresponding Hilbert function.

As in the standard graded case, one would also like to know properties of these points. We quickly point the reader to some of the known results about these sets of points in the literature. The second author studied the minimal generators of these ideals in [95]. Giuffrida, Maggioni, and Ragusa computed the bigraded minimal free resolutions of points in generic position in $\mathbb{P}^1 \times \mathbb{P}^1$ in [37, 38]. The Castelnuovo-Mumford regularity of reduced sets of points in generic position was studied by Hà and the second author in [56]. The regularity of fat points in $\mathbb{P}^1 \times \mathbb{P}^1$ whose support was in generic position was investigated by Sidman and the second author [89]. Finally, Francisco and the second author [27] showed that for small sets of fat points whose support is in generic position, the associated ideal is a monomial ideal that is also componentwise linear.

Chapter 4
Classification of ACM sets of points in $\mathbb{P}^1 \times \mathbb{P}^1$

The coordinate ring of a finite sets of points in $\mathbb{P}^n$ is *always* a Cohen-Macaulay ring. However, the multigraded coordinate ring of a set of points in $\mathbb{P}^1 \times \mathbb{P}^1$, or more generally, a set of points in $\mathbb{P}^{n_1} \times \cdots \times \mathbb{P}^{n_r}$, may fail to have this highly desirable property. This feature is one of the fundamental differences between sets of points in a single projective space and sets of points in a multiprojective space.

It is therefore natural to determine when a set of points is arithmetically Cohen-Macaulay, that is, the associated coordinate ring is Cohen-Macaulay. In this chapter we present one of the main theorems of this monograph, namely, a classification of arithmetically Cohen-Macaulay sets of points of $\mathbb{P}^1 \times \mathbb{P}^1$. In fact, we give five different characterizations of these sets of points. We characterize arithmetically Cohen-Macaulay sets of points in terms of their Hilbert functions, the tuples α_X and β_X, the set S_X, the property $(\star)$, and the separators of X.

We open this chapter by introducing some properties of ACM sets of points. We then proceed to the main result of this chapter. We conclude with a discussion on the open question of characterizing arithmetically Cohen-Macaulay sets of points in other multiprojective spaces.

4.1 Arithmetically Cohen-Macaulay sets of points

We define and describe the relevant properties of arithmetically Cohen-Macaulay (ACM) sets of points in $\mathbb{P}^1 \times \mathbb{P}^1$.

Definition 4.1. Let X be a finite set of distinct points in $\mathbb{P}^1 \times \mathbb{P}^1$. The set X is said to be *arithmetically Cohen-Macaulay* (ACM) if $R/I(X)$ is a Cohen-Macaulay ring.

© The Authors 2015

E. Guardo, A. Van Tuyl, *Arithmetically Cohen-Macaulay Sets of Points in* $\mathbb{P}^1 \times \mathbb{P}^1$,
SpringerBriefs in Mathematics, DOI 10.1007/978-3-319-24166-1_4

To determine if a set of points is ACM, we need to be able to compute both the depth and the Krull dimension of $R/I(X)$. However, as the next lemma shows, the Krull dimension of $R/I(X)$ is the same for all sets of points $X \subseteq \mathbb{P}^1 \times \mathbb{P}^1$, while the depth can only be one of two possible values.

Lemma 4.2. *Let X be a finite set of distinct points in $\mathbb{P}^1 \times \mathbb{P}^1$. Then*

$$1 \leq \text{depth}(R/I(X)) \leq \text{K-dim}(R/I(X)) = 2.$$

Proof. By Lemma 3.5, there is a nonzero-divisor in $R/I(X)$, which implies that $1 \leq \text{depth}(R/I(X))$. To complete the proof, it suffices to prove that the Krull dimension of $R/I(X)$ is two. We view the bigraded ring $R/I(X)$ as $\mathbb{N}^1$-graded (see Remark 2.2), and compute the dimension of this ring.

Suppose $X = \{P_1,\ldots,P_s\}$. By Theorem 3.1, each ideal $I(P_i)$ for $i = 1,\ldots,s$ is a graded ideal generated by two linear polynomials. Furthermore, the generators are linearly independent (it has one generator only in the x_is and another only in the y_is). Thus, as a variety in $\mathbb{P}^3$, $V(I(P_i))$ is a line, and hence $\dim V(I(P_i)) = 1$. The ideal $I(X)$, as a graded ideal, corresponds to a union of lines (see Remark 3.3), hence, $\dim V(I(X)) = 1$, and consequently, $\text{K-dim}(R/I(X)) = \dim V(I(X)) + 1 = 2$.

Remark 4.3. Lemma 4.2 can be generalized. In particular, if X is a finite set of points in $\mathbb{P}^{n_1} \times \cdots \times \mathbb{P}^{n_r}$ and $S = k[\mathbb{P}^{n_1} \times \cdots \times \mathbb{P}^{n_r}]$, then the Krull dimension of $S/I(X)$ is r, the number of projective spaces. So, when X is a finite set of points in $\mathbb{P}^n$, the Krull dimension of $S/I(X)$ is one. Lemma 3.5 can be generalized to show the existence of a nonzero-divisor in $S/I(X)$ for any set of points $X \subseteq \mathbb{P}^{n_1} \times \cdots \times \mathbb{P}^{n_r}$. Note that for sets of points in $\mathbb{P}^n$, we always have $\text{depth}(S/I(X)) = 1 = \text{K-dim}(S/I(X))$, that is, $S/I(X)$ is Cohen-Macaulay.

We pause to make some comments on the bigraded minimal free resolution of $I(X)$. By the Auslander-Buchsbaum Formula (Theorem 2.22), the above result implies that the bigraded minimal free resolution of $R/I(X)$ has either length two or three, and in particular, X is ACM if and only if the length of this resolution is two. As first proved by Giuffrida, Maggioni, and Ragusa [36], the bigraded minimal free resolution of $R/I(X)$ has a number of additional properties that hold for all sets of points $X \subseteq \mathbb{P}^1 \times \mathbb{P}^1$. For the interested reader, we record some of these results, but omit their proof since we will not require them. Note that in Chapter 5 we will explicitly work out the bigraded minimal free resolution of $I(X)$ when X is an ACM set of points.

Theorem 4.4 ([36, Proposition 3.3]). *Let X be a finite set of distinct points in $\mathbb{P}^1 \times \mathbb{P}^1$. Then the bigraded minimal free resolution of $I(X)$ has the form*

$$0 \to \oplus_{i=1}^{p} R(-j_{2,i}, -j'_{2,i}) \oplus_{i=1}^{n} R(-j_{1,i}, -j'_{1,i}) \oplus_{i=1}^{m} R(-j_{0,i}, -j'_{0,i}) \to I(X) \to 0$$

where

(i) $n+1=m+p$.
(ii) $\sum_{i=1}^{p} j_{2,i} - \sum_{i=1}^{n} j_{1,i} + \sum_{i=1}^{m} j_{0,i} = \sum_{i=1}^{p} j'_{2,i} - \sum_{i=1}^{n} j'_{1,i} + \sum_{i=1}^{m} j'0,i = 0$.
(iii) $|X| = -\sum_{i=1}^{m} j_{0,i} j'_{0,i} + \sum_{i=1}^{n} j_{1,i} j'_{1,i} - \sum_{i=1}^{p} j_{2,i} j'_{2,i}$.

As Lemma 4.2 demonstrates, the depth of $R/I(X)$ is either one or two. Both cases can occur, as we show in the next example.

Example 4.5. A set consisting of a single point is ACM. To see this fact, we can make a change of coordinates, and then assume $P = [1:0]\times[1:0]$. Then $I(X) = (x_1, y_1)$, and thus $R/I(X) \cong k[x_0, y_0]$. At this point, we can use the well-known fact that a polynomial ring is Cohen-Macaulay. Alternatively, $\{\overline{x}_0, \overline{y}_0\}$ is a regular sequence modulo $I(X)$ of length two. Since the length of this regular sequence equals K-dim$(R/I(X))$, the set consisting of a single point is ACM.

On the other hand, consider the set of points of Example 3.6. After a change of coordinates, we can take $X = \{[1:0]\times[1:0], [0:1]\times[0:1]\}$. In this case $I(X) = (x_0x_1, x_0y_1, x_1y_0, y_0y_1)$, and as a graded ideal, $I(X)$ is the defining ideal of two skew-lines in $\mathbb{P}^3$, which is a classical example of a non-ACM variety. Alternatively, Giuffrida, Maggioni, and Ragusa [36] show that the depth of $R/I(X)$ is one for this set of points.

When X is an ACM set of points in $\mathbb{P}^1\times\mathbb{P}^1$, we know that depth$(R/I(X))$ must be equal to two, i.e., we can find a regular sequence of length two on $R/I(X)$. In fact, as the next result shows, we can pick this regular sequence so that the elements of the sequence are bihomogeneous of specific degrees. This theorem plays an important role in our classification theorem.

Theorem 4.6. *Let X be a finite set of distinct points in $\mathbb{P}^1\times\mathbb{P}^1$. If X is ACM, then there exist elements $\overline{L}_1, \overline{L}_2$ in $R/I(X)$ such that $L_1 \in R_{1,0}$ and $L_2 \in R_{0,1}$, and L_1, L_2 give rise to a regular sequence in $R/I(X)$.*

Proof. By Lemma 3.5, there exists an $L_1 \in R_{1,0}$ that is a nonzero-divisor on $R/I(X)$. So, it suffices to show that there exists an element of $R_{0,1}$ that is a nonzero-divisor on $R/(I(X), L_1)$.

Because X is ACM, the ring $R/(I(X), L_1)$ is a Cohen-Macaulay ring with K-dim$(R/(I(X), L_1)) = 1$. Suppose that $(I(X), L_1) = Q_1 \cap \cdots \cap Q_t$ is the primary decomposition of $(I(X), L_1)$ with associated primes $\wp_i = \sqrt{Q_i}$. The set of zero-divisors of $R/(I(X), L_1)$ is given by $Z(R/(I(X), L_1)) = \bigcup_{i=1}^{t} \wp_i$.

Suppose that there is no nonzero-divisor of degree $(0,1)$. So

$$R_{0,1} \subseteq Z(R/(I(X), L_1)).$$

Thus $R_{0,1} = \bigcup_{i=1}^{t} (\wp_i)_{0,1}$. But $R_{0,1}$ is an infinite vector space that cannot be written as a union of smaller vector spaces unless there is some i such that $(\wp_i)_{0,1} = R_{0,1}$. On the other hand, by Lemma 3.28, we have $(x_0, x_1) \subseteq \wp_j$ for all j. In particular, if $R_{0,1} \subseteq \wp_i$, then $\mathbf{m} = (x_0, x_1, y_0, y_1) \subseteq \wp_i$, i.e., the maximal ideal $\mathbf{m}$, which has height four in R, is associated with $(I(X), L)$. But this is a contradiction because $R/(I(X), L_1)$ is a Cohen-Macaulay ring of dimension one, so $(I(X), L_1)$ is an unmixed ideal, that is, all associated prime ideals have height three.

Remark 4.7. When I is a homogeneous ideal of $S = k[x_0, \ldots, x_n]$, and S/I is Cohen-Macaulay, then one can always pick a regular sequence of the appropriate length with the property that each element in the regular sequence is homogeneous of degree 1 (e.g., see [12, Proposition 1.5.12]). However, this is no longer true if S/I is multigraded, i.e., we may not be able to find a regular sequence whose elements are all multihomogeneous. For example, let $S = k[x,y]$ with $\deg x = (1,0)$ and $\deg y = (0,1)$ and $I = (xy)$. Then S/I is Cohen-Macaulay, but all the bihomogeneous elements of S/I, which have the form $\overline{cx^a}$ or $\overline{cy^b}$ with $c \in k$, are zero-divisors. Note that $\overline{x} + \overline{y}$ is a nonzero-divisor, but it is not bihomogeneous. Theorem 4.6 is thus a very special situation.

One tool to determine if a set of points in $\mathbb{P}^1 \times \mathbb{P}^1$ is ACM is to use the Auslander-Buchsbaum Formula (Theorem 2.22). Recall that this formula enables us to compute the depth of R/I if we know the projective dimension of R/I. As an example, we use the Auslander-Buchsbaum Formula to show that a set of points on a single ruling in $\mathbb{P}^1 \times \mathbb{P}^1$ is ACM.

Lemma 4.8. *Let $X \subseteq \mathbb{P}^1 \times \mathbb{P}^1$ be a set of points of the form*

$$X = \{A \times B_1, A \times B_2, \ldots, A \times B_v\}.$$

Then X is ACM.

Proof. The defining ideal of X is $I(X) = (H_A, V_{B_1} V_{B_2} \cdots V_{B_v})$. Because $I(X)$ is generated by a regular sequence of length two, the minimal free bigraded resolution of $R/I(X)$ is given by Lemma 2.26:

$$0 \longrightarrow R(-1,-v) \longrightarrow R(-1,0) \oplus R(0,-v) \longrightarrow R \longrightarrow R/I(X) \longrightarrow 0.$$

So, proj-dim$(R/I(X)) = 2$ (recall that when we count the length of the resolution, we start counting with 0 instead of 1). Because K-dim$(R) = 4$, Theorem 2.22 implies that depth$(R/I(X)) = 2$, i.e., X is ACM.

The final result of this section provides some insight into the structure of the ideal of a set of points $X \subseteq \mathbb{P}^1 \times \mathbb{P}^1$ in the case that there exists a horizontal ruling H_A that meets X such that for every $B \in \pi_2(X)$, the point $A \times B$ belongs to X. This theorem will prove to be a very useful tool in our classification result in the next section.

Theorem 4.9. *Let $X \subseteq \mathbb{P}^1 \times \mathbb{P}^1$ be a set of points with $\alpha_X = (\alpha_1, \ldots, \alpha_h)$ and $\beta_X = (\beta_1, \ldots, \beta_v)$. Let H be the degree $(1,0)$ line that contains α_1 points of X, and let $G = V_{B_1} \cdots V_{B_v}$ be the product of the v degree $(0,1)$ lines that contain X. Set $Y = X \cap H$ and $Z = X \setminus Y$. If $\pi_2(Z) \subseteq \pi_2(Y)$, then*

$$I(X) = H \cdot I(Z) + (G).$$

Moreover, we have the following short exact sequence

$$0 \longrightarrow R(-1,-v) \xrightarrow{\phi_2} I(Z)(-1,0) \oplus R(0,-v) \xrightarrow{\phi_1} I(X) = H \cdot I(Z) + (G) \longrightarrow 0$$

where $\phi_1 = [H \;\; G]$ *and* $\phi_2 = \begin{bmatrix} G \\ -H \end{bmatrix}$.

Proof. Because $X = Z \cup Y$, we have $I(X) = I(Z) \cap I(Y)$. The hypothesis $\pi_2(Z) \subseteq \pi_2(Y)$ implies that $\pi_2(Y) = \pi_2(X)$. Indeed, since $Y \subseteq X$, we automatically have $\pi_2(Y) \subseteq \pi_2(X)$. Conversely, if $B \in \pi_2(X)$, then there exists a point $P \in X$ of the form $P = A \times B$. If P lies on the horizontal ruling H, then $P \in Y$, whence $B \in \pi_2(X)$. If P is not on H, then $P \in Z$. But then $B \in \pi_2(Z) \subseteq \pi_2(Y)$.

The ideal $I(Y) = (H,G)$ because Y is the collection of $\alpha_1 = |\pi_2(X)| = v$ points on the horizontal ruling defined by H. In addition, the hypothesis that $\pi_2(Z) \subseteq \pi_2(Y)$ implies that $G \in I(Z)$.

We now show that $I(X) = H \cdot I(Z) + (G)$. Suppose that

$$L = HK_1 + GK_2 \in H \cdot I(Z) + (G)$$

with $K_1 \in I(Z)$ and $K_2 \in R$. Because H and G are in $I(Y)$, we have $L \in I(Y)$. On the other hand, because $K_1, G \in I(Z)$, L is also in $I(Z)$. Hence $H \cdot I(Z) + (G) \subseteq I(Z) \cap I(Y) = I(X)$.

Conversely, let $L \in I(Z) \cap I(Y)$. Because $L \in I(Y)$, $L = HK_1 + GK_2$. By our choice of H, H vanishes at all points whose first coordinate is A_1. If we can show that $K_1 \in I(Z)$, then we will have completed the proof. Now because $G \in I(Z)$, we also have $HK_1 \in I(Z)$. But for every $A \times B \in Z$ with $A \neq A_1$, then $H(A \times B) \neq 0$. Hence $HK_1 \in I(Z)$ if and only if $K_1(A \times B) = 0$ for every $A \times B \in Z$.

We now prove the statement about the short exact sequence. Because $I(Y) = (H,G)$ where $\deg H = (1,0)$ and $\deg G = (0,v)$, the ideal $I(Y)$ is a complete intersection. By Lemma 2.26 the bigraded minimal free resolution of $I(Y)$ is

$$0 \longrightarrow R(-1,-v) \xrightarrow{\phi_2} R(-1,0) \oplus R(0,-v) \xrightarrow{\phi_1} I(Y) = (H,G) \longrightarrow 0$$

where $\phi_1 = [H \;\; G]$ and $\phi_2 = \begin{bmatrix} G \\ -H \end{bmatrix}$. We note that for every $L \in R(-1,-v)$, we have $\phi_2(L) = (GL, -HL)$. But because $G \in I(Z)$, we in fact have $\operatorname{im} \phi_2 \subseteq I(Z)(-1,0) \oplus R(0,-r)$. This fact, coupled with the fact that $I(Z) = H \cdot I(Z) + (G)$, gives us the following short exact sequence of graded R-modules:

$$0 \longrightarrow R(-1,-v) \xrightarrow{\phi_2} I(Z)(-1,0) \oplus R(0,-v) \xrightarrow{\phi_1} I(X) = H \cdot I(Z) + (G) \longrightarrow 0$$

where ϕ_1 and ϕ_2 are the same as the maps above.

Remark 4.10. For readers familiar with the concept of linkage, the above result is an example of a Basic Double Link. See the notes of Migliore and Nagel [73, Lemma 6] for more details about linkage.

4.2 Classification of ACM sets of points

We are now ready to give a classification of the arithmetically Cohen-Macaulay sets of points in $\mathbb{P}^1 \times \mathbb{P}^1$. As we see in the statement below, there are five equivalent ways to classify these points.

Theorem 4.11 (Classification of ACM sets of points in $\mathbb{P}^1 \times \mathbb{P}^1$). *Let X be a finite set of distinct points in $\mathbb{P}^1 \times \mathbb{P}^1$. Then the following are equivalent:*

(i) *X is an ACM set of points.*
(ii) *ΔH_X is the bigraded Hilbert function of an artinian quotient of $k[x_1, y_1]$.*
(iii) *$\alpha_X^* = \beta_X$.*
(iv) *S_X has no incomparable elements with respect to the partial order $\succeq$ on $\mathbb{N}^v$.*
(v) *X satisfies property $(\star)$ (see Definition 3.19).*
(vi) *For every $P \in X$, $|\deg_X(P)| = 1$.*

Proof. The equivalence of statements (iii), (iv), and (v) is Theorem 3.21.

$(i) \Rightarrow (ii)$ Because $R/I(X)$ is ACM, by Lemma 4.6 there is an $L_1 \in R_{1,0}$ and an $L_2 \in R_{0,1}$ such that $\overline{L}_1, \overline{L}_2$ is a regular sequence on $R/I(X)$. After a linear change of variables in the x_is and in the y_js, we can assume that $L_1 = x_0$ and $L_2 = y_0$.

We now have two short exact sequences of degree $(0,0)$:

$$0 \longrightarrow (R/I(X))(-1,0) \xrightarrow{\times \overline{x}_0} R/I(X) \longrightarrow R/(I(X), x_0) \longrightarrow 0$$

and

$$0 \longrightarrow R/(I(X), x_0))(0,-1) \xrightarrow{\times \overline{y}_0} R/(I(X), x_0) \longrightarrow R/(I(X), x_0, y_0) \longrightarrow 0.$$

From these two short exact sequences, we can compute the Hilbert function of $R/(I(X), x_0, y_0)$. In particular

$$\begin{aligned} H_{R/(I(X),x_0,y_0)}(i,j) &= H_{R/(I(X),x_0)}(i,j) - H_{R/(I(X),x_0)}(i,j-1) \\ &= (H_X(i,j) - H_X(i-1,j)) - (H_X(i,j-1) - H_X(i-1,j-1)) \\ &= \Delta H_X(i,j). \end{aligned}$$

That is, the Hilbert function of $R/(I(X), x_0, y_0)$ is given by the first difference function ΔH_X.

To complete the proof, it suffices to show that

$$R/(I(X),x_0,y_0) \cong \frac{R/(x_0,y_0)}{(I(X),x_0,y_0)/(x_0,y_0)}$$

is an artinian quotient of $k[x_1,y_1]$. It is immediate that $R/(x_0,y_0) \cong k[x_1,y_1]$, with $\deg x_1 = (1,0)$ and $\deg y_1 = (0,1)$. Moreover, by Lemma 3.5, we have $(x_0,x_1)^h \subseteq (I(X),x_0)$ and $(y_0,y_1)^v \subseteq (I(X),y_0)$. If $t = \max\{h,v\}$, then

$$(x_0^t,x_1^t,y_0^t,y_1^t) \in (I(X),x_0,y_0).$$

In other words, $(I(X),x_0,y_0)/(x_0,y_0)$ is an artinian ideal of $k[x_1,y_1]$.

(ii) $\Rightarrow$ *(iii)* Suppose ΔH_X is the bigraded Hilbert function of an artinian quotient of $k[x_1,y_1]$. By Theorem 2.29, ΔH_X has the form

$\Delta H_X =$

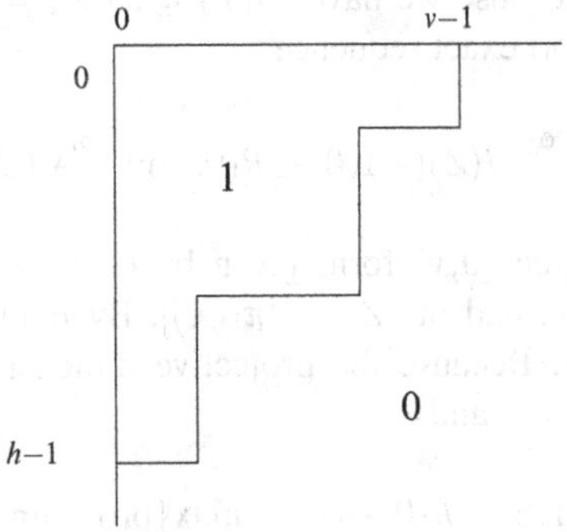

We can recover the Hilbert function H_X from ΔH_X using the identity

$$H_X(i,j) = \sum_{(0,0)\preceq(k,l)\preceq(i,j)} \Delta H_X(k,l).$$

Comparing this formula to Theorem 3.29, we have

$$\begin{aligned}
H_X(h-1,0) &= h = \alpha_1^* &&= \Delta H_X(0,0)+\Delta H_X(1,0)+\cdots+\Delta H_X(h-1,0)\\
H_X(h-1,1) &= \alpha_1^*+\alpha_2^* &&= \alpha_1^*+\Delta H_X(0,1)+\Delta H_X(1,1)+\cdots+\Delta H_X(h-1,1)\\
H_X(h-1,2) &= \alpha_1^*+\alpha_2^*+\alpha_3^* &&= \alpha_1^*+\alpha_2^*+\Delta H_X(0,2)+\Delta H_X(1,2)+\cdots+\Delta H_X(h-1,2)\\
&\vdots &&\vdots
\end{aligned}$$

In other words,

$$\alpha_j^* = \Delta H_X(0,j-1)+\cdots+\Delta H_X(h-1,j-1)$$

counts the number of 1's that appears in the $(j-1)$-th column of ΔH_X. By a similar argument, β_j^* counts the number of 1's that appears in the $(j-1)$-th row of ΔH_X.

Because ΔH_X consists of only 1's and 0's, we can view ΔH_X as a Ferrers diagram. But then it follows that $(\alpha_1^*, \ldots, \alpha_{\alpha_1}^*)^* = \beta_X^*$, that is, $\alpha_X = \beta_X^*$.

$(v) \Rightarrow (i)$ We prove the statement by induction on $|\pi_1(X)|$. If $|\pi_1(X)| = 1$, then X is ACM by Lemma 4.8.

Now suppose that $|\pi_1(X)| > 1$. We first claim that there is a point $A \in \pi_1(X)$ such that $\pi_2(X \cap H_A) = \pi_2(X)$. We always have $\pi_2(X \cap H_A) \subseteq \pi_2(X)$. Let $A \in \pi_1(X)$ with $|\pi_2(X \cap H_A)|$ maximal. In other words, $|\pi_2(X \cap H_A)| = \alpha_1$. We will show that the point A is the desired point.

Suppose that there is a point $B \in \pi_2(X) \backslash \pi_2(X \cap H_A)$. So, there exists an $A \neq A' \in \pi_1(X)$ such that $A' \times B \in X$. Let $B' \in \pi_2(X \cap H_A)$ be any point. So $A \times B'$ and $A' \times B$ are points in X. By the hypotheses, $A \times B$ or $A' \times B'$ are in X. But $A \times B \notin X$ (else $B \in \pi_2(X \cap H_A)$). So, for each $B' \in \pi_2(X \cap H_A)$, $A' \times B' \in X$. But this means $|\pi_2(X \cap H_{A'})| > |\pi_2(X \cap H_A)|$, contradicting the maximality of $|\pi_2(X \cap H_A)|$. This now proves the claim.

For the induction step, set $Y = X \setminus H_A$, where A is the point from the above claim and $Z = X \cap H_A$. Because we have $\pi_2(Y) \subseteq \pi_2(Z) = \pi_2(X)$, Theorem 4.9 implies that we have the short exact sequence

$$0 \longrightarrow R(-1,-v) \xrightarrow{\phi_2} I(Z)(-1,0) \oplus R(0,-v) \xrightarrow{\phi_1} I(X) = H_A \cdot I(Z) + (G) \longrightarrow 0.$$

Here, G is the degree $(0,v)$ form given by $G = V_{B_1} \cdots V_{B_v}$. Now the set Z still satisfies property $(\star)$ and $|\pi_1(Z)| < |\pi_1(X)|$. By induction, Z is ACM, and hence proj-dim$(I(Z)) = 1$. Because the projective dimension of a free R-module is 0, proj-dim$(R(-1,v)) = 0$ and

$$\begin{aligned}\text{proj-dim}(I(Z)(-1,0) \oplus R(0,-v)) &= \max\{\text{proj-dim}(I(Z)), \text{proj-dim}(R(0,-v)\} \\ &= \text{proj-dim}(I(Z)).\end{aligned}$$

Using the above short exact sequence and Lemma 2.23 we have proj-dim$(I(X)) \leq 1$. Because $I(X)$ is not a free R-module, we thus have proj-dim$(I(X)) = 1$. But then depth$(R/I(X)) = 2$ by the Auslander-Buchsbaum formula (Theorem 2.22). Thus, X is ACM as desired.

$(i) \Rightarrow (vi)$ By Theorem 4.6, after a change of coordinates, we can assume that x_0 and y_0 form a regular sequence on $R/I(X)$.

Suppose for a contradiction that there is a point $P \in X$ with $t = |\deg_X(P)| > 1$. Let F and G be the two smallest minimal separators with respect to the lexicographical order (i.e., order the t minimal separators of P by degree according to the lexicographical order, and consider the two smallest such separators). If $\deg F = (a_1, a_2)$ and $\deg G = (b_1, b_2)$, then because F and G are minimal separators, we cannot have $\deg F \preceq \deg G$. So $a_1 < b_1$ and $a_2 > b_2$.

Let $\underline{c} = (b_1, a_2)$. Since $(a_1, a_2) \preceq (b_1, a_2)$, by Theorem 3.39 we must have that $\dim_k (I(X), F)_{\underline{c}} = \dim_k I(X)_{\underline{c}} + 1$. So, a basis for $(I(X), F)_{\underline{c}}$ is given by the $\dim_k I(X)_{\underline{c}}$ basis elements of $I(X)_{\underline{c}}$ and any form of degree $\underline{c}$ in $(I(X), F)_{\underline{c}} \backslash I(X)_{\underline{c}}$. One such form is $x_0^{b_1 - a_1} F$. A similar argument implies that $\dim_k (I(X), G)_{\underline{c}} = \dim_k I(X)_{\underline{c}} + 1$, so a basis for $(I(X),G)_{\underline{c}}$ is given by the $\dim_k I(X)_{\underline{c}}$ basis elements of $I(X)_{\underline{c}}$ and $y_0^{a_2-b_2}G$. Note that we are using the fact that x_0 and y_0 form a regular sequence on $R/I(X)$, so they do not vanish at any of the points of X.

Since $\deg G \preceq \underline{c}$, and $\deg F \preceq \underline{c}$ we have

$$(I(X),F)_{\underline{c}} \subseteq I(Y)_{\underline{c}} \text{ and } (I(X),G)_{\underline{c}} \subseteq I(Y)_{\underline{c}}.$$

But $\dim_k(I(X),F)_{\underline{c}} = \dim_k I(X)_{\underline{c}} + 1$, and since $\dim_k I(Y)_{\underline{c}} \leq \dim_k I(X)_{\underline{c}} + 1$, we must have $(I(X),F)_{\underline{c}} = I(Y)_{\underline{c}}$. A similar argument implies that $(I(X),G)_{\underline{c}} = I(Y)_{\underline{c}}$. Hence,

$$(I(X),F)_{\underline{c}} = (I(X),G)_{\underline{c}}.$$

Because $y_0^{a_2-b_2}G \in (I(X),G)_{\underline{c}}$, our discussion about the basis for $(I(X),F)_{\underline{c}}$ implies

$$y_0^{a_2-b_2}G = H + cx_0^{b_1-a_1}F \text{ with } H \in I(X)_{\underline{c}} \text{ and } 0 \neq c \in k.$$

Note that $c \neq 0$ because if $c = 0$, then the right-hand side vanishes at all the points of X, but the left-hand side does not because G and y_0 do not vanish at P. We thus have

$$y_0^{a_2-b_2}G - cx_0^{b_1-a_1}F \in I(X).$$

Because x_0 and y_0 form a regular sequence on $R/I(X)$ and since X is ACM, y_0 is a nonzero-divisor of $R/(I(X),x_0)$. But then, since $y_0^{a_2-b_2}G \in (I(X),x_0)$, and y_0 is a nonzero-divisor, we must have $G \in (I(X),x_0)$, that is,

$$G = H_1 + H_2x_0 \text{ with } H_1 \in I(X) \text{ and } H_2 \in R.$$

If $Q \in X\backslash\{P\}$, then $G(Q) = 0$ implies that $H_2(Q) = 0$ since $H_1(Q) = 0$ and $x_0(Q) \neq 0$. On the other hand, if we evaluate both sides at P, we have

$$0 \neq G(P) = H_1(P) + x_0(P)H_2(P) = x_0(P)H_2(P).$$

But because $x_0(P) \neq 0$, this forces $H_2(P) \neq 0$. So, H_2 is a separator of P with $\deg H_2 = (b_1 - 1, b_2)$. So, let K be a minimal separator with $\deg K \preceq \deg H_2$. But then $\deg K <_{\text{lex}} \deg G = (b_1, b_2)$. But any minimal separator whose degree is smaller than G with respect to the lexicographical order must have the same degree as F, i.e., $\deg F = \deg K$. So, $\deg F \preceq \deg H_2 = (b_1 - 1, b_2)$. But this contradicts the fact that $a_2 > b_2$ and hence $\deg F \npreceq \deg H_2$. This gives the desired contradiction.

(vi) $\Rightarrow$ *(i)* We will prove the contrapositive statement: if X is not ACM, then there exists a point $P \in X$ such that $|\deg_X(P)| > 1$.

Let $\pi_1(X) = \{A_1, \ldots, A_h\}$ and $\pi_2(X) = \{B_1, \ldots, B_v\}$ be the set of first coordinates, respectively, second coordinates, that appear in X. Because we have already proved the equivalence of statements (i) and (v), we know that if X is not ACM, then there exist points $A \times B$ and $A' \times B'$ in X such that $A \times B'$ and $A' \times B$ are not in X. After relabelling, we can assume $A \times B = A_1 \times B_1$ and $A' \times B' = A_2 \times B_2$.

We set $X_{A_1} = \{A \times B \in X \mid A = A_1\}$ and $X_{B_1} = \{A \times B \in X \mid B = B_1\}$. We thus have

$$\begin{aligned} X_{A_1} &= \{A_1 \times B_1, A_1 \times B_{i_2}, \ldots, A_1 \times B_{i_b}\} = X \cap H_{A_1} \\ X_{B_1} &= \{A_1 \times B_1, A_{j_2} \times B_1, \ldots, A_{j_a} \times B_1\} = X \cap V_{B_1}. \end{aligned}$$

Note that $A_1 \times B_2 \notin X_{A_1}$ and $A_2 \times B_1 \notin X_{B_1}$.

There are now four cases to consider. In each case we show $|\deg_X(A_1 \times B_1)| > 1$.

Case 1: $|X_{A_1}| = |X_{B_1}| = 1$.

In this case $A_1 \times B_1$ is the only point of X with first coordinate A_1 and second coordinate B_1. The two forms $F_1 = H_{A_2}H_{A_3}\cdots H_{A_h}$ and $F_2 = V_{B_2}V_{B_3}\cdots V_{B_v}$ are separators of $A_1 \times B_1$ of degrees $(h-1,0)$ and $(0,v-1)$, respectively. (It is not hard to see that F_1 and F_2 pass through all the points of $X \setminus \{A_1 \times B_1\}$.) If $|\deg_X(A_1 \times B_1)| = 1$, then there would be a minimal separator F of $A_1 \times B_1$ such that $\deg F \preceq (h-1,0)$ and $\deg F \preceq (0,v-1)$. But this would mean that $\deg F = (0,0)$. However, there is no separator of $A_1 \times B_1$ of degree $(0,0)$. Thus $|\deg_X(A_1 \times B_1)| > 1$.

Case 2: $|X_{A_1}| > 1$ and $|X_{B_1}| = 1$.

The two forms $F_1 = H_{A_2}H_{A_3}\cdots H_{A_h}V_{B_{i_2}}V_{B_{i_3}}\cdots V_{B_{i_b}}$ and $F_2 = V_{B_2}V_{B_3}\cdots V_{B_v}$ are two separators of $A_1 \times B_1$ in X. If $|\deg_X(A_1 \times B_1)| = 1$, then there would exist a minimal separator F such that $\deg F \preceq (h-1,b-1)$ and $\deg F \preceq (0,v-1)$, or in other words, $\deg F \preceq (0,b-1)$. Note that F would also be a separator of $A_1 \times B_1$ in X_{A_1}, and thus, by Lemma 3.41, we will have $(0,b-1) \preceq \deg F$, and thus $\deg F = (0,b-1)$.

Because $\deg_{X_{A_1}}(A_1 \times B_1) = \{(0,b-1)\}$, by Theorem 3.38 F is the unique (up to scalar multiplication in $R/I(X)$) separator of $A_1 \times B_1$ in X_{A_1}. Now because $V_{B_{i_2}}V_{B_{i_3}}\cdots V_{B_{i_b}}$ is another separator of degree $(0,b-1)$ for the point $A_1 \times B_1$ in X_{A_1}, we have $F = cV_{B_{i_2}}V_{B_{i_3}}\cdots B_{B_{i_b}} + H$ for some nonzero scalar c and $H \in I(X)$. However, it then follows that $F(A_2 \times B_2) \neq 0$, contradicting the fact that F is a separator of $A_1 \times B_1$ in X. So, $|\deg_X(P)| > 1$.

Case 3: $|X_{A_1}| = 1$ and $|X_{B_1}| > 1$.

The proof is similar to the previous case.

Case 4: $|X_{A_1}| > 1$ and $|X_{B_1}| > 1$.

The forms

$$F_1 = H_{A_2}H_{A_3}\cdots H_{A_h}V_{B_{i_2}}V_{B_{i_3}}\cdots V_{B_{i_b}} \text{ and } F_2 = H_{A_{j_2}}H_{A_{j_3}}\cdots H_{A_{j_a}}V_{B_2}V_{B_3}\cdots V_{B_v}$$

are two separators of $A_1 \times B_1$ of degrees $(h-1,b-1)$ and $(a-1,v-1)$, respectively. If $|\deg_X(A_1 \times B_1)| = 1$, then there would exist a separator F of

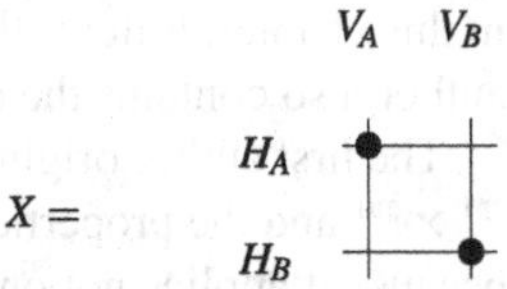

Fig. 4.1 Two non-collinear points

$A_1 \times B_1$ with $\deg F \preceq (h-1, b-1)$ and $\deg F \preceq (a-1, v-1)$. In other words, $\deg F \preceq (a-1, b-1)$. Now such an F would also be a separator of $A_1 \times B_1$ in the set of points $X' = X_{A_1} \cup X_{B_1}$. But then by Theorem 3.41, this would mean $(a-1, b-1) \preceq \deg F$. Thus $\deg F = (a-1, b-1)$.

By Theorem 3.41, the point $A_1 \times B_1$ in the set of points X' has $\deg_{X'}(A_1 \times B_1) = \{(a-1, b-1)\}$, and thus by Theorem 3.38, the form F must be the unique (up to scalar multiplication in $R/I(X)$) separator of $A_1 \times B_1$ in X'. On the other hand, the form $F' = H_{A_{j_2}} H_{A_{j_3}} \cdots H_{A_{j_a}} V_{B_{i_2}} V_{B_{i_3}} \cdots V_{B_{i_b}}$ is also a separator of degree $(a-1, b-1)$ of $A_1 \times B_1$ in X', and thus $F = cF' + H$ for some c and $H \in I(X)$. But then $F(A_2 \times B_2) \neq 0$, contradicting the fact that F must pass through every point of $X \setminus \{A_1 \times B_1\}$. Hence, we must have $|\deg_X(A_1 \times B_1)| > 1$.

Example 4.12. Conditions (*iv*) and (*v*) of Theorem 4.11 make it easy to "see" if a set of points is ACM. For example, consider the set of points of Example 3.6 which we have redrawn in Figure 4.1. Since we see that that this configuration does not satisfy property ($\star$), it cannot be ACM. It can be also seen computing S_X. We have $S_X = \{(1,0),(0,1)\}$ and the two elements are incomparable with respect to the partial order $\succeq$ on $\mathbb{N}^2$. So, X cannot be ACM.

The tuples α_X and β_X provide an easy way to determine if a set of points is ACM. For example, for the points of Example 3.10, we saw that $\alpha_X = (3,2,2)$ and $\beta_X = (2,2,2,1)$. Because $\alpha_X^* = (3,3,1) \neq \beta_X$, the set of points cannot be ACM.

On the other hand, the points of Example 3.16 are ACM because they satisfy property ($\star$). Alternatively, we have $\alpha_X = (6,5,3,1,1)$ Since $\alpha_X^* = (5,3,3,2,2,1) = \beta_X$, the set of points is ACM. We can also see that X is ACM from the set S_X. In this case the set

$$S_X = \{(1,1,1,1,1,1),(1,1,1,1,1,0),(1,1,1,0,0,0),(1,0,0,0,0,0),(1,0,0,0,0,0)\}$$

has no incomparable elements with respect to the partial order $\succeq$ on $\mathbb{N}^6$.

4.3 Additional notes

Giuffrida, Maggioni, and Ragusa [36] gave the first characterization of ACM sets of points in $\mathbb{P}^1 \times \mathbb{P}^1$. As in Theorem 4.11, their classification is in terms of the first difference function of H_X. Although they do not explicitly state that ΔH_X is the Hilbert function of an artinian quotient of $k[x_1, y_1]$, this fact can be deduced from their work. The connection to bigraded artinian quotients, and the proof presented in this chapter, follows the work of the second author [94]. The paper of the second author also contains the equivalent statement about the vectors α_X and β_X.

The first author originally proved the equivalence between ACM sets of points in $\mathbb{P}^1 \times \mathbb{P}^1$ and the properties of S_X (see [42, 43]). Her result, however, is more general because it applies not only to reduced points, but also to fat points. We will turn to the classification of ACM sets of fat points in $\mathbb{P}^1 \times \mathbb{P}^1$ in Chapter 6.

The connection between ACM sets of points and the degree of a separator was first discovered by Marino [71]. The proof presented here uses a shorter proof discovered by the authors [50].

The work of this chapter suggests the following natural problem:

Problem 4.13. *Find a classification of ACM sets of points in* $\mathbb{P}^{n_1} \times \cdots \times \mathbb{P}^{n_r}$.

All sets of points in $\mathbb{P}^n$ are ACM (see Remark 4.7). The work of this chapter provides a solution to Problem 4.13 for sets of points in $\mathbb{P}^1 \times \mathbb{P}^1$. For no other multiprojective space $\mathbb{P}^{n_1} \times \cdots \times \mathbb{P}^{n_r}$ is a solution known.

In an attempt to solve Problem 4.13, the authors considered the natural generalizations of some of the statements of Theorem 4.11 in the paper [48]. Surprisingly, none of the equivalent statements appear to be generalizable. For example, it was shown that one could not classify ACM sets of points in $\mathbb{P}^2 \times \mathbb{P}^2$ by their Hilbert functions. The authors found two sets of points, X_1 and X_2, such that the Hilbert functions were equal, i.e., $H_{X_1} = H_{X_2}$, but one set was ACM while the other was not.

The geometric property ($\star$) was also not generalizable. While it was true that points in $\mathbb{P}^1 \times \mathbb{P}^n$ that satisfy property ($\star$) are ACM, there also exist ACM sets of points in $\mathbb{P}^1 \times \mathbb{P}^n$ that fail to satisfy property ($\star$).

One direction of Theorem 4.11 continues to hold more generally for separators. In particular, if X is an ACM set of points in $\mathbb{P}^{n_1} \times \cdots \times \mathbb{P}^{n_r}$, then $|\deg_X(P)| = 1$ for all $P \in X$. However, the converse fails to hold; an example exists of points in $\mathbb{P}^2 \times \mathbb{P}^2$ such that every point $P \in X$ has $|\deg_X(P)| = 1$, but X is not ACM.

It is possible, (and we expect), that Theorem 4.11 can be extended to points in $\mathbb{P}^1 \times \cdots \times \mathbb{P}^1$ (r times). The first open case is to classify ACM sets of points X in $\mathbb{P}^1 \times \mathbb{P}^1 \times \mathbb{P}^1$. If $I(X)$ is the associated defining ideal in $S = k[\mathbb{P}^1 \times \mathbb{P}^1 \times \mathbb{P}^1]$, then we know that $\operatorname{depth}(S/I(X))$ is 1, 2, or 3. When the depth is three, then X is ACM, and we know that ΔH_X is the Hilbert function of a trigraded artinian quotient of $k[x_1, y_1, z_1]$. In [48], it is shown that if $\operatorname{depth}(S/I(X)) = 2$, then ΔH_X cannot be the Hilbert function of a trigraded artinian quotient. To complete the classification, one would need to also show that if the depth of $S/I(X)$ is one, then ΔH_X cannot have the Hilbert function of a trigraded artinian quotient of $k[x_1, y_1, z_1]$.

Chapter 5
Homological invariants

As we saw in the last chapter, we can determine if a set of points X in $\mathbb{P}^1 \times \mathbb{P}^1$ is ACM directly from a combinatorial description of the points. In this chapter we show that this combinatorial description, in particular, the tuples α_X and β_X, also allows us to determine the bigraded Betti numbers in the bigraded minimal free resolution of $I(X)$ when X is ACM. Consequently, the Hilbert function of X when X is ACM can also be computed directly from α_X and β_X or from the set S_X. We conclude this chapter by answering the interpolation question introduced in Chapter 1. Specifically, we classify what functions $H : \mathbb{N}^2 \to \mathbb{N}$ are the Hilbert functions of ACM reduced sets of points in $\mathbb{P}^1 \times \mathbb{P}^1$.

5.1 Bigraded minimal free resolution

In this section we study the bigraded minimal free resolution of the ideal $I(X)$ when X is an ACM set of points in $\mathbb{P}^1 \times \mathbb{P}^1$. (For some general results about the bigraded minimal free resolution for any set of points X, see Theorem 4.4.) We start with a lemma about a special configuration of points.

Lemma 5.1. *Let X be a set of $s = hv$ points in $\mathbb{P}^1 \times \mathbb{P}^1$ such that*

$$\alpha_X = (\underbrace{v,\ldots,v}_{h}) \text{ and } \beta_X = (\underbrace{h,\ldots,h}_{v}).$$

Then X is ACM. In fact, $I(X)$ is a complete intersection, and the bigraded minimal free resolution of $I(X)$ is

$$0 \longrightarrow R(-h,-v) \longrightarrow R(-h,0) \oplus R(0,-v) \longrightarrow I(X) \longrightarrow 0.$$

Proof. The set X is ACM by Theorem 4.11 because $\alpha_X^* = \beta_X$.

© The Authors 2015

E. Guardo, A. Van Tuyl, *Arithmetically Cohen-Macaulay Sets of Points in* $\mathbb{P}^1 \times \mathbb{P}^1$,
SpringerBriefs in Mathematics, DOI 10.1007/978-3-319-24166-1_5

Since $|\alpha_X| = h$ and $|\beta_X| = v$, it follows that $\pi_1(X) = \{A_1,\dots,A_h\}$ and $\pi_2(X) = \{B_1,\dots,B_v\}$ where $A_i, B_j \in \mathbb{P}^1$. Because $|X| = hv$, the set X must be the set of points

$$X = \{A_i \times B_j \mid 1 \le i \le h, 1 \le j \le v\}.$$

Hence, if $I(A_i \times B_j) = (H_{A_i}, V_{B_j})$ is the bihomogeneous prime ideal associated with the point $A_i \times B_j$, then

$$\begin{aligned} I(X) &= \bigcap_{\substack{1\le i\le h\\ 1\le j\le v}} \left(H_{A_i}, V_{B_j}\right) = \bigcap_{1\le i\le h} (H_{A_i}, V_{B_1}V_{B_2}\cdots V_{B_v}) \\ &= (H_{A_1}H_{A_2}\cdots H_{A_h}, V_{B_1}V_{B_2}\cdots V_{B_v}). \end{aligned}$$

Since $\deg H_{A_1}H_{A_2}\cdots H_{A_h} = (h,0)$ and $\deg V_{B_1}V_{B_2}\cdots V_{B_v} = (0,v)$, the two generators of $I(X)$ form a regular sequence on R, and hence, $I(X)$ is a complete intersection. The bigraded minimal free resolution then follows from Lemma 2.26.

We can restate Lemma 5.1 in terms of S_X.

Lemma 5.2. *Let X be a set of $s = hv$ points in $\mathbb{P}^1 \times \mathbb{P}^1$ such that*

$$S_X = \{s_1,\dots,s_h\} \text{ where } s_i = (\underbrace{1,\dots,1}_{v}) \text{ for all } i = 1,\dots,h.$$

Then X is ACM, and $I(X)$ is a complete intersection.

Proof. The set S_X has no incomparable elements with respect to the partial ordering $\succeq$ on $\mathbb{N}^v$. So X is ACM by Theorem 4.11. In this case $\alpha_X = (\underbrace{v,\dots,v}_{h})$ and $\beta_X = (\underbrace{h,\dots,h}_{v})$, so the fact that $I(X)$ is a complete intersection follows from Lemma 5.1.

To state our result about the resolution of $I(X)$, we require the following notation. Suppose that $X \subseteq \mathbb{P}^1 \times \mathbb{P}^1$ is an ACM set of points with $\alpha_X = (\alpha_1,\dots,\alpha_h)$. Set

$$C_X := \{(h,0),(0,\alpha_1)\} \cup \{(i-1,\alpha_i) \mid \alpha_i - \alpha_{i-1} < 0\},$$

and

$$V_X := \{(h,\alpha_h)\} \cup \{(i-1,\alpha_{i-1}) \mid \alpha_i - \alpha_{i-1} < 0\}.$$

We take $\alpha_{-1} = 0$. With this notation, we can state the main result of this section.

Theorem 5.3. *Suppose that X is an ACM set of points in $\mathbb{P}^1 \times \mathbb{P}^1$ with $\alpha_X = (\alpha_1,\dots,\alpha_h)$. Let C_X and V_X be constructed from α_X as above. Then the bigraded minimal free resolution of $I(X)$ has the form*

$$0 \longrightarrow \bigoplus_{(v_1,v_2)\in V_X} R(-v_1,-v_2) \longrightarrow \bigoplus_{(c_1,c_2)\in C_X} R(-c_1,-c_2) \longrightarrow I(X) \longrightarrow 0.$$

Proof. We proceed by induction on the tuple $(|\pi_1(X)|,|X|)$. If v is any positive integer, and $(|\pi_1(X),|X|)=(1,v)$, then $\alpha_X=(v)$ and $\beta_X=(\underbrace{1,\ldots,1}_{v})$, i.e., X consists of v points on one horizontal ruling. In this case, X is also a complete intersection, so by Lemma 5.1, the bigraded minimal free resolution of $I(X)$ has the form

$$0 \longrightarrow R(-1,-v) \longrightarrow R(-1,0)\oplus R(0,-v) \longrightarrow I(X) \longrightarrow 0.$$

From α_X, we have $C_X=\{(1,0),(0,v)\}$ and $V_X=\{(1,v)\}$; the bigraded shifts then agree with the formula in the statement of the theorem.

So, suppose $(|\pi_1(X),|X|)=(h,s)$ and the theorem holds for all $Y\subseteq \mathbb{P}^1\times\mathbb{P}^1$ with $(|\pi_1(Y)|,|Y|)<_{lex}(h,s)$. Suppose that $\alpha_X=(\underbrace{\alpha_1,\ldots,\alpha_1}_{l},\alpha_{l+1},\ldots,\alpha_h)$, i.e., $\alpha_{l+1}<\alpha_1$, but $\alpha_1=\alpha_2=\cdots=\alpha_l$.

If $l=h$, then X is a complete intersection and the resolution is given by Lemma 5.1. The conclusion now follows because $C_X=\{(h,0),(0,\alpha_h)\}$ and $V_X=\{(h,\alpha_h)\}$.

So, suppose that $l<h$ and suppose that $A_1,\ldots,A_l$ are the l points of $\pi_1(X)$ that have $|\pi_1^{-1}(A_i)\cap X|=\alpha_1$. Set $Y=(\pi_1^{-1}(A_1)\cup\cdots\cup\pi_1^{-1}(A_l))\cap X$. Because X is ACM, we have $\alpha_X=\beta_X^*$, and in particular, $\alpha_1=|\beta_X|=v$. Hence,

$$Y=\{A_i\times B_j \mid 1\le i\le l,\ B_j\in\pi_2(X)\}$$

with $\alpha_Y=(\underbrace{\alpha_1,\ldots,\alpha_1}_{l})$ and $\beta_Y=(\underbrace{l,\ldots,l}_{\alpha_1})$. By Lemma 5.1, Y is a complete intersection. Furthermore, if we set $F=H_{A_1}\cdots H_{A_l}$ and $G=V_{B_1}\cdots V_{B_v}$, then again by Lemma 5.1 we have

$$0 \longrightarrow R(-l,-\alpha_1) \xrightarrow{\phi_2} R(-l,0)\oplus R(0,-\alpha_1) \xrightarrow{\phi_1} I(Y) \longrightarrow 0 \qquad (\dagger)$$

where $\phi_1=[F\ \ G]$ and $\phi_2=\begin{bmatrix} G\\ -F\end{bmatrix}$. (Recall that $v=\alpha_1$.)

Let $Z=X\backslash Y$. Because $\pi_2(Z)\subseteq\pi_2(X)$, it follows that $G=V_{B_1}\cdots V_{B_v}\in I(Z)$. Hence, $\operatorname{im}\phi_2\subseteq I(Z)(-l,0)\oplus R(0,-\alpha_1)$ since $v=\alpha_1$. We now prove the following claim.

Claim. $I(X)=F\cdot I(Z)+(G)$.

Proof of the Claim. By construction, $X=Z\cup Y$, and thus $I(X)=I(Z)\cap I(Y)$. Hence, we want to show that $I(Z)\cap I(Y)=F\cdot I(Z)+(G)$.

So, if $K\in F\cdot I(Z)+(G)$, then there is an $H_1\in I(Z)$ and an $H_2\in R$ such that $K=FH_1+GH_2$. Because $F,G\in I(Y)$ and $H_1,G\in I(Z)$, so we have $K\in I(Z)\cap I(Y)$.

To show the reverse inclusion, let $K \in I(Z) \cap I(Y)$. Since $K \in I(Y)$, there exist $H_1, H_2 \in R$ such that $K = FH_1 + GH_2$. It now suffices to show that $H_1 \in I(Z)$. Take any point $A \times B \in Z$. By construction, $A \neq A_i$ for $i = 1, \ldots, l$. Now

$$0 = K(A \times B) = F(A \times B)H_1(A \times B) + G(A \times B)H_2(A \times B).$$

Since $G \in I(X) \subseteq I(Z)$, we have $G(A \times B) = 0$ and so $F(A \times B)H_1(A \times B) = 0$. But since $F = H_{A_1} \cdots H_{A_l}$, it cannot vanish at $A \times B$, so $H_1(A \times B) = 0$. So $H_1 \in I(Z)$ as desired. □

From the resolution for $I(Y)$ given in (†), the claim, and the fact that

$$\operatorname{im} \phi_2 \subseteq I(Z)(-l, 0) \oplus R(0, -\alpha_1),$$

we have the following short exact sequence of R-modules

$$0 \longrightarrow R(-l, -\alpha_1) \xrightarrow{\phi_2} I(Z)(-l, 0) \oplus R(0, -\alpha_1) \xrightarrow{\phi_1} I(X) = F \cdot I(Z) + (G) \longrightarrow 0$$

where ϕ_1 and ϕ_2 are as above.

We now first observe that Z is ACM because it satisfies property $(\star)$. Indeed, let $A_i \times B_j$ and $A_r \times B_s$ be in Z with $i \neq r$ and $j \neq s$. Since $Z \subseteq X$, and because X is ACM, by Theorem 4.11, either $A_i \times B_s$ or $A_r \times B_j$ is in X. Now Z is formed by removing all the points from X whose first coordinate is in the set $\{A_1, \ldots, A_l\}$. Since $A_i \times B_j$ and $A_r \times B_s$ are in Z, this means that neither A_i nor A_r belong to $\{A_1, \ldots, A_l\}$. So, the point $A_i \times B_s$ or $A_r \times B_j$ would belong to Z, meaning Z is ACM. Note that by the construction of Z, we have $\alpha_Z = (\alpha_{l+1}, \ldots, \alpha_t)$.

Because the induction hypothesis holds for Z, we can use the above short exact sequence and the mapping cone construction to construct a bigraded resolution for $I(X)$. In particular, we get

$$0 \longrightarrow \left[\bigoplus_{(v_1, v_2) \in V_Z} R(-(v_1 + l), -v_2) \right] \oplus R(-l, -\alpha_1) \longrightarrow$$

$$\left[\bigoplus_{(c_1, c_2) \in C_Z} R(-(c_1 + l), -c_2) \right] \oplus R(0, -\alpha_1) \longrightarrow I(X) \longrightarrow 0.$$

Because the resolution has length 2, and because X is ACM, the resolution of $I(X)$ cannot be made shorter by the Auslander-Buchsbaum formula (cf. Theorem 2.22).

To show that this resolution is minimal, it is enough to show that no tuple in the set

$$\{(c_1 + l, c_2) \mid (c_1, c_2) \in C_Z\} \cup \{(0, \alpha_1)\}$$

is in the set

$$\{(v_1+l,v_2) \mid (v_1,v_2) \in V_Z\} \cup \{(l,\alpha_1)\}.$$

That is, no further cancellation of the Betti numbers is allowed. By the induction hypothesis, we can assume that no $(c_1,c_2) \in C_Z$ is in V_Z, and hence

$$\{(c_1+l,c_2) \mid (c_1,c_2) \in C_Z\} \cap \{(v_1+l,v_2) \mid (v_1,v_2) \in V_Z\} = \emptyset.$$

If $(c_1+l,c_2) = (l,\alpha_1)$ for some $(c_1,c_2) \in C_Z$, then this implies that $(0,\alpha_1) \in C_Z$. But this contradicts the induction hypothesis because the only element in C_Z of the form $(0,*)$ is $(0,\alpha_{l+1})$ and $\alpha_{l+1} < \alpha_1$. Similarly, if $(0,\alpha_1) \in \{(v_1+l,v_2) \mid (v_1,v_2) \in V_Z\}$, this implies $(-l,\alpha_1) \in V_Z$, which is again a contradiction of the induction hypothesis. So the resolution given above is also minimal.

To complete the proof we only need to verify that

(*i*) $C_X = \{(c_1+l,c_2) \mid (c_1,c_2) \in C_Z\} \cup \{(0,\alpha_1)\}$.
(*ii*) $V_X = \{(v_1+l,v_2) \mid (v_1,v_2) \in V_Z\} \cup \{(l,\alpha_1)\}$.

We now work through the requisite details.

Let $C' = \{(c_1+l,c_2) \mid (c_1,c_2) \in C_Z\} \cup \{(0,\alpha_1)\}$. By definition

$$C_X = \{(h,0),(0,\alpha_1)\} \cup \{(i-1,\alpha_i) \mid \alpha_i - \alpha_{i-1} < 0\}$$

and

$$C_Z = \{(h-l,0),(0,\alpha_{l+1})\} \cup \{(k-1,\alpha_{l+k}) \mid \alpha_{l+k} - \alpha_{l+k-1} < 0\}$$

because $\alpha_X = (\underbrace{\alpha_1,\ldots,\alpha_1}_{l},\alpha_{l+1},\ldots,\alpha_h)$ and $\alpha_Z = (\alpha_{l+1},\ldots,\alpha_h)$.

We check that $C_X \subseteq C'$. It is immediate that the elements $(0,\alpha_1)$ and $(h,0)$ are in C'. So, suppose $(d_1,d_2) \in \{(i-1,\alpha_i) \mid \alpha_i - \alpha_{i-1} < 0\}$. For $i \leq l$, $\alpha_i = \alpha_1$. So, if $(d_1,d_2) \in \{(i-1,\alpha_i) \mid \alpha_i - \alpha_{i-1} < 0\}$, then $(d_1,d_2) = (l,\alpha_{l+1})$, or there exists some positive k such that $(d_1,d_2) = (l+k-1,\alpha_{l+k})$ and $\alpha_{l+k} - \alpha_{l+k-1} < 0$. But in either case, $(d_1-l,d_2) \in C_Z$, and hence, $(d_1,d_2) \in C'$.

Conversely, we only need to check that $(c_1+l,c_2) \in C_X$ for every $(c_1,c_2) \in C_Z$. It is straightforward to check that $((h-l)+l,0) \in C_X$. Also, as noted, $(l,\alpha_{l+1}) \in C_X$. If $(c_1,c_2) \in \{(k-1,\alpha_{l+k}) \mid \alpha_{l+k} - \alpha_{l+k-1} < 0\}$, then $(c_1+l,c_2) = (k+l-1,\alpha_{l+k}) \in C_Z$ because $\alpha_{l+k} - \alpha_{l+k-1} < 0$. Thus $C_X = C'$.

The proof of (*ii*) is similar in nature. Again, for completeness, we will verify the details. Let $V' = \{(v_1+l,v_2) \mid (v_1,v_2) \in V_Z\} \cup \{(l,\alpha_1)\}$. By definition,

$$V_X = \{(h,\alpha_h)\} \cup \{(i-1,\alpha_{i-1}) \mid \alpha_i - \alpha_{i-1} < 0\}$$

since $\alpha_X = (\underbrace{\alpha_1,\ldots,\alpha_1}_{l},\alpha_{l+1},\ldots,\alpha_h)$, and

$$V_Z = \{(h-l,\alpha_h)\} \cup \{(k-1,\alpha_{k+l-1}) \mid \alpha_{l+k} - \alpha_{l+k-1} < 0\}$$

because $\alpha_Z = (\alpha_{l+1},\ldots,\alpha_h)$.

We first check that $V_X \subseteq V'$. The element $(h,\alpha_h) \in V'$ because $(h-l,\alpha_h) \in V_Z$. So, suppose $(d_1,d_2) \in \{(i-1,\alpha_{i-1}) \mid \alpha_i - \alpha_{i-1} < 0\}$. Because $\alpha_i = \alpha_1$ for $1 \leq i \leq l$, it is either the case that $(d_1,d_2) = (l,\alpha_l) = (l,\alpha_1)$ or $(d_1,d_2) = (l+k-1,\alpha_{l+k-1})$ with $k > 1$. But in the first case, it is immediate that $(d_1,d_2) \in V'$. In the second case, because $(d_1,d_2) \in V_X$, $\alpha_{l+k} - \alpha_{l+k-1} < 0$. But then $(k-1,\alpha_{l+k-1}) \in V_Z$, and hence, $(d_1,d_2) \in V'$.

Conversely, because $(l,\alpha_1) \in V_X$, we only need to check that $(v_1+l,v_2) \in V_X$ for all $(v_1,v_2) \in V_Z$. It is immediate that $(h-l+l,\alpha_h) \in V_X$. So, suppose that $(v_1,v_2) \in \{(k-1,\alpha_{l+k-1}) \mid \alpha_{l+k} - \alpha_{l+k-1} < 0\}$. But then $(v_1+l,v_2) \in \{(l+k-1,\alpha_{l+k-1}) \mid \alpha_{l+k} - \alpha_{l+k-1} < 0\}$. Because $\alpha_X = (\underbrace{\alpha_1,\ldots,\alpha_1}_{l},\alpha_{l+1},\ldots,\alpha_h)$, we must have $\{(l+k-1,\alpha_{l+k-1}) \mid \alpha_{l+k} - \alpha_{l+k-1} < 0\} = V_X \backslash \{(h,\alpha_h)\}$, which completes the proof.

Remark 5.4. Within the above proof is a result similar to Theorem 4.9. We divided the set of points X into two smaller sets of points Y and Z, and related the ideals of $I(Y)$ and $I(Z)$ to the ideal of $I(X)$. Our proof is almost identical to the proof of Theorem 4.9, except in the above proof we are removing a complete intersection of points from X. In fact, when $l = 1$ in the above proof, i.e., if $\alpha_x = (\alpha_1,\alpha_2,\ldots,\alpha_h)$ with $\alpha_1 > \alpha_2$, then we could appeal directly to Theorem 4.9.

Remark 5.5. The resolution of an ACM set of points in $\mathbb{P}^1 \times \mathbb{P}^1$ was first computed by Giuffrida *et al.* [36, Theorem 4.1]. Giuffrida *et al.* showed that the bigraded Betti numbers for an ACM set of points $X \subseteq \mathbb{P}^1 \times \mathbb{P}^1$ could be determined from the first difference function ΔH_X. Because X is ACM, ΔH_X is the Hilbert function of a bigraded artinian quotient of $k[x_1,y_1]$ by Theorem 4.11, and thus, has the form An

$\Delta H_X =$

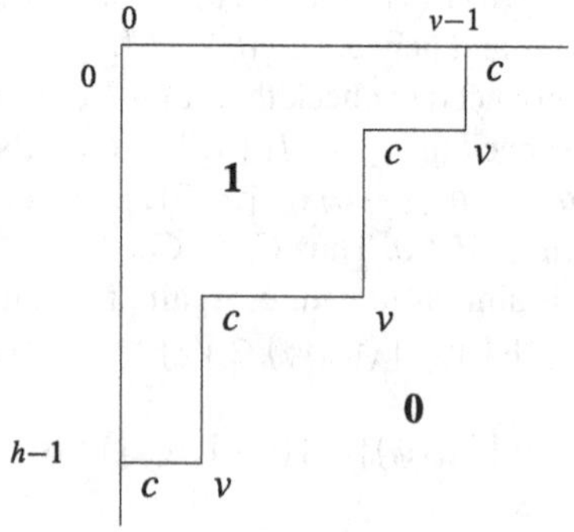

element of C_X, which they called a *corner* of ΔH_X, corresponds to a tuple (i,j) that is either $(h,0),(0,\alpha_1)=(0,v)$ or has the property that $\Delta H_X(i,j)=0$, but $\Delta H_X(i-1,j)=\Delta H_X(i,j-1)=1$. We have labelled the corners of ΔH_X with a c in the above diagram. An element of V_X is a *vertex*. A tuple (i,j) is called a vertex if $\Delta H_X(i,j)=\Delta H_X(i-1,j)=\Delta H_X(i,j-1)=0$, but $\Delta H_X(i-1,j-1)=1$. We have labelled the vertices of ΔH_X with a v in the above diagram. Our proof shows that the required information is encoded directly into the tuple α_X.

As a final corollary, we observe that we can explicitly write out a set of minimal generators of $I(X)$ when X is ACM.

Corollary 5.6. *Suppose that X is an ACM set of points in $\mathbb{P}^1\times\mathbb{P}^1$ with $\alpha_X=(\alpha_1,\ldots,\alpha_h)$. Let $H_{A_1},\ldots,H_{A_h}$ denote the associated horizontal rulings and let $V_{B_1},\ldots,V_{B_v}$ denote the associated vertical rulings which minimally contain X. Then a minimal homogeneous set of generators of $I(X)$ is given by*

$$\{H_{A_1}\cdots H_{A_h},V_{B_1}\cdots V_{B_v}\}\cup\{H_{A_1}\cdots H_{A_i}V_{B_1}\cdots V_{B_{\alpha_{i+1}}}\mid \alpha_{i+1}-\alpha_i<0\}.$$

Proof. Each element in the above list passes through all the points of X, so they belong to $I(X)$. Furthermore, the bidegrees of these forms agree with the bidegrees of the generators of $I(X)$ as given in Theorem 5.3.

5.2 Hilbert function

Given a set of ACM points X in $\mathbb{P}^1\times\mathbb{P}^1$, Theorem 5.3 enables one to compute the bigraded minimal free resolution, and consequently, the Hilbert function of X directly from the tuple α_X. While one could deduce the following result directly from the resolution of of $I(X)$ given in Theorem 5.3, the details become somewhat messy. Instead, we give a more direct proof.

Theorem 5.7. *Suppose that X is an ACM set of points in $\mathbb{P}^1\times\mathbb{P}^1$ with $\alpha_X=(\alpha_1,\ldots,\alpha_h)$. Then*

$$H_X=\begin{bmatrix}1&2&\cdots&\alpha_1-1&\alpha_1&\alpha_1&\cdots\\1&2&\cdots&\alpha_1-1&\alpha_1&\alpha_1&\cdots\\\vdots&\vdots&&\vdots&\vdots&\vdots&\ddots\end{bmatrix}+\begin{bmatrix}0&0&\cdots&0&0&0&\cdots\\1&2&\cdots&\alpha_2-1&\alpha_2&\alpha_2&\cdots\\1&2&\cdots&\alpha_2-1&\alpha_2&\alpha_2&\cdots\\\vdots&\vdots&&\vdots&\vdots&\vdots&\ddots\end{bmatrix}+$$

$$\begin{bmatrix}0&0&\cdots&0&0&0&\cdots\\0&0&\cdots&0&0&0&\cdots\\1&2&\cdots&\alpha_3-1&\alpha_3&\alpha_3&\cdots\\1&2&\cdots&\alpha_3-1&\alpha_3&\alpha_3&\cdots\\\vdots&\vdots&&\vdots&\vdots&\vdots&\ddots\end{bmatrix}+\cdots+\begin{bmatrix}0&0&\cdots&0&0&0&\cdots\\\vdots&\vdots&&\vdots&\vdots&\vdots&\\0&0&\cdots&0&0&0&\cdots\\1&2&\cdots&\alpha_h-1&\alpha_h&\alpha_h&\cdots\\1&2&\cdots&\alpha_h-1&\alpha_h&\alpha_h&\cdots\\\vdots&\vdots&&\vdots&\vdots&\vdots&\ddots\end{bmatrix}.$$

Proof. Our proof is by induction on the tuple $(|\pi_1(X)|, |X|)$. For any $v \in \mathbb{N}$, if $(|\pi_1(X), |X|) = (1, v)$, then $\alpha_X = (v)$ and $\beta_x = (\underbrace{1, \dots, 1}_{1})$. The Hilbert function of X, which is given by Lemma 3.26, has the form

$$H_X = \begin{bmatrix} 1 & 2 & 3 & \cdots & v-1 & v & v & \cdots \\ 1 & 2 & 3 & \cdots & v-1 & v & v & \cdots \\ \vdots & \vdots & \vdots & & \vdots & \vdots & \ddots \end{bmatrix},$$

and H_X agrees with the statement of the theorem.

Now suppose that $(|\pi_1(X)|, |X|) = (h, s)$ and that the theorem holds for all ACM sets of points $Z \subseteq \mathbb{P}^1 \times \mathbb{P}^1$ with $(|\pi_1(Z)|, |Z|) <_{lex} (h, s)$.

Let H_A be the $(1,0)$ line that contains α_1 points of X. Set $Y = X \cap H_A$ and $Z = X \setminus Y$. Because X is ACM, we must have $\pi_2(Z) \subseteq \pi_2(Y)$ (the proof of this fact is just as in Theorem 4.11). So by Theorem 4.9, we have a short exact sequence

$$0 \longrightarrow R(-1, -\alpha_1) \longrightarrow I(Z)(-1,0) \oplus R(0, -\alpha_1) \longrightarrow I(X) \longrightarrow 0.$$

From this short exact sequence, we deduce that

$$\begin{aligned} H_X(i,j) &= \dim_k R_{i,j} - \dim_k I(X)_{i,j} \\ &= \dim_k R_{i,j} - [\dim_k I(Z)_{i-1,j} + \dim_k R_{i,j-\alpha_1} - \dim_k R_{i-1,j-\alpha_1}] \end{aligned}$$

where $\dim_k R_{a,b} = 0$ if $(0,0) \not\preceq (a,b)$.

Let $n_+ := \max\{0, n\}$. Now $\dim_k R_{i,j} = (i+1)(j+1)$, $\dim_k R_{i,j-\alpha_1} = (i+1)(j-\alpha_1+1)_+$ and $\dim_k R_{i-1,j-\alpha_1} = i(j-\alpha_1+1)_+$. Thus,

$$\begin{aligned} \dim_k R_{i,j} - \dim_k R_{i,j-\alpha_1} + \dim_k R_{i-1,j-\alpha_1} &= (i+1)(j+1) - (i+1)(j-\alpha_1+1)_+ + i(j-\alpha_1+1)_+ \\ &= i(j+1) + (j+1) - (j-\alpha_1+1)_+ \\ &= \dim_k R_{i-1,j} + H_Y(i,j) \end{aligned}$$

where the last equality follows from Lemma 3.26 since $H_Y(i,j) = \min\{j+1, \alpha_1\} = (j+1) - (j-\alpha_1+1)_+$. Hence

$$H_X(i,j) = H_Y(i,j) + \dim_k R_{i-1,j} - \dim_k I(Z)_{i-1,j} = H_Y(i,j) + H_Z(i-1,j).$$

The set of points Z is ACM because it satisfies property $(\star)$. Furthermore, by construction, $\alpha_Z = (\alpha_2, \dots, \alpha_h)$, so we can apply the induction hypothesis to evaluate H_Z. Therefore, if we use Lemma 3.26 to compute H_Y and apply induction to compute H_Z, we have

$$H_X = \begin{bmatrix} 1 & 2 & \cdots & \alpha_1 - 1 & \alpha_1 & \alpha_1 & \cdots \\ 1 & 2 & \cdots & \alpha_1 - 1 & \alpha_1 & \alpha_1 & \cdots \\ \vdots & \vdots & \vdots & & \vdots & \vdots & \ddots \end{bmatrix} + \begin{bmatrix} 0 & 0 & \cdots & 0 & 0 & 0 & \cdots \\ 1 & 2 & \cdots & \alpha_2 - 1 & \alpha_2 & \alpha_2 & \cdots \\ 1 & 2 & \cdots & \alpha_2 - 1 & \alpha_2 & \alpha_2 & \cdots \\ \vdots & \vdots & & \vdots & \vdots & \vdots & \ddots \end{bmatrix}$$

$$+ \cdots + \begin{bmatrix} 0 & 0 & \cdots & 0 & 0 & 0 & \cdots \\ \vdots & \vdots & & \vdots & \vdots & \vdots & \\ 0 & 0 & \cdots & 0 & 0 & 0 & \cdots \\ 1 & 2 & \cdots & \alpha_h - 1 & \alpha_h & \alpha_h & \cdots \\ 1 & 2 & \cdots & \alpha_h - 1 & \alpha_h & \alpha_h & \cdots \\ \vdots & \vdots & & \vdots & \vdots & \vdots & \ddots \end{bmatrix}.$$

As a corollary, we observe that ΔH_X, the first difference function of H_X can be written down directly from the tuple α_X or from S_X when X is ACM. In fact, ΔH_X has the shape of the Ferrers diagram of α_X.

Corollary 5.8. *Suppose that X is an ACM set of points in $\mathbb{P}^1 \times \mathbb{P}^1$ with $\alpha_X = (\alpha_1, \ldots, \alpha_h)$. Then*

$$\Delta H_X = \begin{bmatrix} \underbrace{1 \; 1 \; \cdots \; 1}_{\alpha_1} \; 1 \; 0 \; 0 \; \cdots \\ \underbrace{1 \; \cdots \; 1}_{\alpha_2} \; 1 \; 0 \; 0 \; 0 \; \cdots \\ \vdots \; \vdots \; \vdots \; \vdots \; \vdots \; \vdots \; \vdots \; \cdots \\ \underbrace{1 \; \cdots \; 1}_{\alpha_h} \; 0 \; 0 \; 0 \; 0 \; \cdots \\ 0 \; 0 \; 0 \; 0 \; 0 \; 0 \; 0 \; \cdots \\ \vdots \; \vdots \; \vdots \; \vdots \; \vdots \; \vdots \; \vdots \; \ddots \end{bmatrix}.$$

Proof. The result follows directly from Theorem 5.7 once we observe that the first difference of the matrix

$$\begin{bmatrix} 0 & 0 & \cdots & 0 & 0 & 0 & \cdots \\ \vdots & \vdots & & \vdots & \vdots & \vdots & \\ 0 & 0 & \cdots & 0 & 0 & 0 & \cdots \\ 1 & 2 & \cdots & \alpha_i - 1 & \alpha_i & \alpha_i & \cdots \\ 1 & 2 & \cdots & \alpha_i - 1 & \alpha_i & \alpha_i & \cdots \\ \vdots & \vdots & & \vdots & \vdots & \vdots & \ddots \end{bmatrix}$$

is given by

$$\begin{bmatrix} 0\,0 \cdots 0\,0\,0 \cdots \\ \vdots\ \vdots \quad \vdots\ \vdots\ \vdots \\ 0\,0 \cdots 0\,0\,0 \cdots \\ \underbrace{1\,1 \cdots 1\,1}_{\alpha_i}\,0 \cdots \\ 0\,0 \cdots 0\,0\,0 \cdots \\ \vdots\ \vdots \quad \vdots\ \vdots\ \vdots\ \ddots \end{bmatrix}.$$

The first difference function of an ACM set of points in $\mathbb{P}^1 \times \mathbb{P}^1$ is also encoded directly into the set S_X. The following lemma, which is straightforward to verify, captures the relationship between α_X and S_X when X is ACM.

Lemma 5.9. *Suppose that X is an ACM set of points in $\mathbb{P}^1 \times \mathbb{P}^1$ with tuple $\alpha_X = (\alpha_1, \ldots, \alpha_h)$. Set $v = \alpha_1$. Then $S_X = \{s_1, \ldots, s_h\}$ where*

$$s_i = (\underbrace{1, \ldots, 1}_{\alpha_i}, \underbrace{0, \ldots, 0}_{v-\alpha_i}) \text{ for } i = 1, \ldots, h.$$

We can then restate Corollary 5.8 in terms of S_X.

Corollary 5.10. *Suppose that X is an ACM set of points in $\mathbb{P}^1 \times \mathbb{P}^1$ with $S_X = \{s_1, \ldots, s_h\}$ where $s_i \in \mathbb{N}^v$ for $i = 1, \ldots, h$. Then*

$$\Delta H_X = \begin{bmatrix} s_1 & 0 & \cdots \\ s_2 & 0 & \cdots \\ s_3 & 0 & \cdots \\ \vdots & & \\ s_h & 0 & \cdots \\ \mathbf{0} & 0 & \cdots \\ \vdots & \vdots & \ddots \end{bmatrix}.$$

where $\mathbf{0}$ is the $1 \times v$ zero matrix and we view each s_i as an $1 \times v$ matrix.

Example 5.11. Theorem 3.40 describes the Hilbert function of the set of points $X_{h,v}$ without the assumption that it is an ACM set of points. However, from the construction of $X = X_{h,v}$, we have $\alpha_X = (v, \underbrace{1, \ldots, 1}_{h-1})$ and $\beta_X = (h, \underbrace{1, \ldots, 1}_{v-1})$, and we can easily check that $\alpha_X^* = \beta_X$. Hence, X is an ACM set of points by Theorem 4.11. Thus, by Corollary 5.8 or Corollary 5.10, its first difference Hilbert function has α_i's "1"s in the $(i-1)$-th row for all $i \geq 1$, that is,

$$\Delta H_X = \begin{bmatrix} \underbrace{1\ \ 1 \cdots 1}_{\alpha_1 = v}\ 1\ 0\ 0 \cdots \\ \\ \underbrace{1}_{\alpha_2=1}\ \ 0 \cdots 0\ 0\ 0\ 0\ \ 0\ \cdots \\ \\ \vdots\ \ \vdots\ \vdots\ \vdots\ \vdots\ \vdots\ \vdots\ \cdots \\ \underbrace{1}_{\alpha_h=1}\ \ 0 \cdots 0\ 0\ 0\ 0 \cdots \\ \\ 0\ \ 0\ 0\ 0\ 0\ 0\ 0 \cdots \\ \vdots\ \ \vdots\ \vdots\ \vdots\ \vdots\ \vdots\ \vdots\ \ddots \end{bmatrix}.$$

The Hilbert function of $X_{h,v}$ described in Theorem 3.40 can then be recovered from the above function.

5.3 A worked out example

To help the reader understand the results of the previous sections we return to the set of points in Example 3.16. We compute both its bigraded minimal free resolution and its Hilbert function from the associated partition $\alpha_X = (6,5,3,1,1)$. For the convenience of the reader, we have redrawn this set of points in Figure 5.1.

From the figure we can see that $\alpha_X = (6,5,3,1,1)$ and $\beta_X = (5,3,3,2,2,1)$. In particular, $\alpha_X = \beta_X^*$, and so by Theorem 4.11, this set of points is ACM. Alternatively, we can also see that X is ACM from the set S_X:

$$S_X = \{(\underbrace{1,1,1,1,1,1}_{\alpha_1=6}), (\underbrace{1,1,1,1,1}_{\alpha_2=5},0), (\underbrace{1,1,1}_{\alpha_3=3},0,0,0), (\underbrace{1}_{\alpha_4=1},0,0,0,0,0),$$
$$(\underbrace{1}_{\alpha_5=1},0,0,0,0,0)\}.$$

$X =$

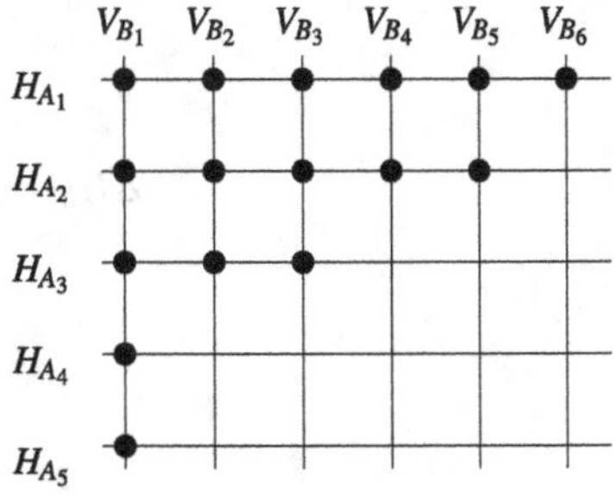

Fig. 5.1 The set of points resembling partition $(6,5,3,1,1)$

We can see that S_X has no incomparable elements with respect to the partial order $\succeq$ on $\mathbb{N}^6$.

We now compute the bigraded minimal free resolution of $I(X)$. To do so, we determine the sets C_X and V_X. We have

$$C_X = \{(5,0),(6,0),(1,5),(2,3),(3,1)\}$$
$$V_X = \{(5,1),(1,6),(2,5),(3,3)\}.$$

So the bigraded minimal free resolution of $I(X)$ has the form

$$0 \longrightarrow \begin{matrix} R(-5,-1)\oplus R(-1,-6) \\ \oplus \\ R(-2,-5)\oplus R(-3,-3) \end{matrix} \longrightarrow \begin{matrix} R(-5,0)\oplus R(-6,0) \\ \oplus R(-1,-5)\oplus \\ R(-2,-3)\oplus R(-3,-1) \end{matrix} \longrightarrow I(X) \longrightarrow 0.$$

Using Theorem 5.7, the bigraded Hilbert function can be computed directly from $\alpha_X = (6,5,3,1,1)$. Specifically,

$$H_X = \begin{bmatrix} 1\,2\,3\,4\,5\,6\,6 & \cdots \\ 1\,2\,3\,4\,5\,6\,6 & \cdots \\ \vdots & \ddots \end{bmatrix} + \begin{bmatrix} 0\,0\,0\,0\,0\,0\,0 & \cdots \\ 1\,2\,3\,4\,5\,5\,5 & \cdots \\ 1\,2\,3\,4\,5\,5\,5 & \cdots \\ \vdots & \ddots \end{bmatrix} + \begin{bmatrix} 0\,0\,0\,0\,0\,0\,0 & \cdots \\ 0\,0\,0\,0\,0\,0\,0 & \cdots \\ 1\,2\,3\,3\,3\,3\,3 & \cdots \\ 1\,2\,3\,3\,3\,3\,3 & \cdots \\ \vdots & \ddots \end{bmatrix} +$$

$$\begin{bmatrix} 0\,0\,0\,0\,0\,0\,0 & \cdots \\ 0\,0\,0\,0\,0\,0\,0 & \cdots \\ 0\,0\,0\,0\,0\,0\,0 & \cdots \\ 1\,1\,1\,1\,1\,1\,1 & \cdots \\ 1\,1\,1\,1\,1\,1\,1 & \cdots \\ \vdots & \ddots \end{bmatrix} + \begin{bmatrix} 0\,0\,0\,0\,0\,0\,0 & \cdots \\ 0\,0\,0\,0\,0\,0\,0 & \cdots \\ 0\,0\,0\,0\,0\,0\,0 & \cdots \\ 0\,0\,0\,0\,0\,0\,0 & \cdots \\ 1\,1\,1\,1\,1\,1\,1 & \cdots \\ 1\,1\,1\,1\,1\,1\,1 & \cdots \\ \vdots & \ddots \end{bmatrix} = \begin{bmatrix} 1 & 2 & 3 & 4 & 5 & 6 & 6 & \cdots \\ 2 & 4 & 6 & 8 & 10 & 11 & 11 & \cdots \\ 3 & 6 & 9 & 11 & 13 & 14 & 14 & \cdots \\ 4 & 7 & 10 & 12 & 14 & 15 & 15 & \cdots \\ 5 & 8 & 11 & 13 & 15 & 16 & 16 & \cdots \\ 5 & 8 & 11 & 13 & 15 & 16 & 16 & \cdots \\ \vdots & \vdots & \vdots & \vdots & \vdots & \vdots & \vdots & \ddots \end{bmatrix}.$$

Moreover, by Corollary 5.8 or Corollary 5.10, the first difference function of H_X is given by

$$\Delta H_X = \begin{bmatrix} 1\,1\,1\,1\,1\,1\,0 & \cdots \\ 1\,1\,1\,1\,1\,0\,0 & \cdots \\ 1\,1\,1\,0\,0\,0\,0 & \cdots \\ 1\,0\,0\,0\,0\,0\,0 & \cdots \\ 1\,0\,0\,0\,0\,0\,0 & \cdots \\ 0\,0\,0\,0\,0\,0\,0 & \cdots \\ \vdots & \ddots \end{bmatrix}.$$

If we only consider the 1's in ΔH_X, then the ones form a Ferrers diagram of $(6,5,3,1,1)$, i.e., ΔH_X encodes the tuple α_X, or the set S_X, when X is ACM. Using the language of corners and vertices of Giuffrida, Maggioni, and Ragusa (see [36] and Remark 5.5), we see that the corners are $\{(0,6),(1,4),(2,3),(3,1),(5,0)\}$ and the vertices are $\{(1,6),(2,4),(3,3),(5,1)\}$, which agrees with what we found above.

5.4 Complete intersection sets of points

As a short application we can classify when the defining ideal of a set of points in $\mathbb{P}^1 \times \mathbb{P}^1$ is a complete intersection. We introduce the following definition.

Definition 5.12. A set of points $X \subseteq \mathbb{P}^1 \times \mathbb{P}^1$ is a *complete intersection* if $I(X)$ is a complete intersection.

Recall that by Lemma 2.25, a complete intersection is Cohen-Macaulay.

Theorem 5.13. *Let $X \subseteq \mathbb{P}^1 \times \mathbb{P}^1$ be a set of points. Then the following are equivalent:*

(*i*) *X is a complete intersection.*
(*ii*) $\alpha_X = (\underbrace{v,\ldots,v}_{h})$ *and* $\beta_X = \alpha_X^* = (\underbrace{h,\ldots,h}_{v})$.
(*iii*) $S_X = \{s_1,\ldots,s_h\}$ *with* $s_i = (\underbrace{1,\ldots,1}_{v})$ *for all* $i = 1,\ldots,h$.

Proof. (*i*) $\Rightarrow$ (*ii*) Suppose that X is a complete intersection. By Lemma 2.25 it follows that X is Cohen-Macaulay, so $\alpha_X^* = \beta_X$ by Theorem 4.11. If $\alpha_X \neq (\underbrace{v,\ldots,v}_{h})$, i.e., if there is some index $i > 1$ such that

$$\alpha_X = (\underbrace{v,\ldots,v}_{i-1},\alpha_i,\ldots)$$

with $v > \alpha_i$, then by Corollary 5.6, the ideal $I(X)$ has at least three generators. But because $I(X)$ is a complete intersection generated by a regular sequence of length at least three, Lemma 2.25 implies that $\dim R/I(X) \leq 4-3 = 1$. However, this contradicts Lemma 4.2 which states that $\dim R/I(X) = 2$ for all sets of points X in $\mathbb{P}^1 \times \mathbb{P}^1$.

(*ii*) $\Rightarrow$ (*iii*) Because $\beta_X = \alpha_X^*$, by Theorem 4.11, the set X is ACM. The description of S_X then follows from Lemma 5.9.

(*iii*) $\Rightarrow$ (*i*) This result is Lemma 5.2.

In light of the above result, a complete intersection can be described in terms of the data v and h. We introduce the following notation for future use.

Fig. 5.2 The complete intersection $CI(4,6)$

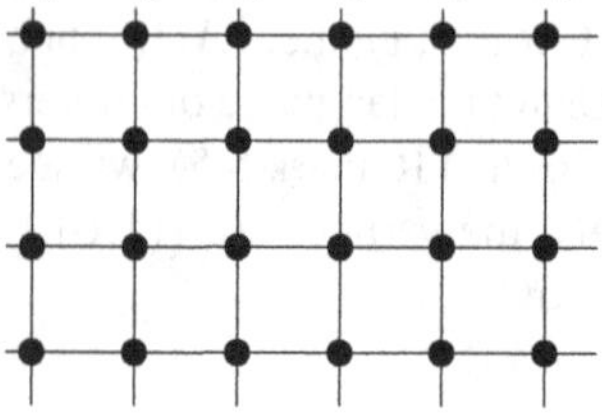

Notation 5.14. Suppose that X is a complete intersection with $\alpha_X = (v,\dots,v)$ and $\beta_X = (h,\dots,h)$, or with $S_X = \{s_1,\dots,s_h\}$ where $s_i = \underbrace{(1,\dots,1)}_{v}$ for all $i = 1,\dots,h$. Then we denote it by $CI(v,h)$.

Example 5.15. We can visualize a complete intersection $X = CI(4,6)$ as in Figure 5.2 where the dots represent the points in X. Here, $h = 4$ and $v = 6$. We also note that $\alpha_X = (6,6,6,6)$ and $\beta_X = (4,4,4,4,4,4)$, or

$$S_X = \{(1,1,1,1,1,1),(1,1,1,1,1,1),(1,1,1,1,1,1),(1,1,1,1,1,1)\}.$$

5.5 Interpolation problem

Recall from Chapter 1 that the interpolation problem (Problem 1.2) asks what can be the Hilbert function of a set of points X in $\mathbb{P}^{n_1} \times \cdots \times \mathbb{P}^{n_r}$. Except when $r = 1$, this problem remains wide open. In fact, for $r > 1$, the only known complete solution is for the case of ACM sets of points in $\mathbb{P}^1 \times \cdots \times \mathbb{P}^1$. (Classifying the Hilbert functions of *all* sets of points in $\mathbb{P}^1 \times \mathbb{P}^1$ remains an open problem.) In this section, we present the classification of Hilbert functions of ACM sets of points in $\mathbb{P}^1 \times \mathbb{P}^1$.

Recall that if $H : \mathbb{N}^2 \to \mathbb{N}$ is a numerical function, we define

$$\Delta H(i,j) = H(i,j) - H(i-1,j) - H(i,j-1) + H(i-1,j-1) \text{ for all } (i,j) \in \mathbb{N}^2$$

with the convention that $H(a,b) = 0$ if $(a,b) \not\succeq (0,0)$.

Theorem 5.16. *Let $H : \mathbb{N}^2 \to \mathbb{N}$ be a numerical function. Then H is the Hilbert function of an ACM set of points X in $\mathbb{P}^1 \times \mathbb{P}^1$ if and only ΔH is the bigraded Hilbert function of an artinian quotient of $k[x_1,y_1]$.*

Proof. Suppose that $H = H_X$ is the Hilbert function of an ACM set of points X in $\mathbb{P}^1 \times \mathbb{P}^1$. By Theorem 4.11, ΔH_X must be the Hilbert function of a bigraded artinian quotient of $k[x_1,y_1]$, thus proving one direction of the statement.

For the converse direction, suppose that ΔH is the Hilbert function of an artinian quotient of $k[x_1,y_1]$. We now need to construct an ACM set of points X in $\mathbb{P}^1 \times \mathbb{P}^1$ such that $\Delta H_X = \Delta H$.

It follows from Theorem 2.29 that ΔH must have the form. Let α_i denote the

$\Delta H =$

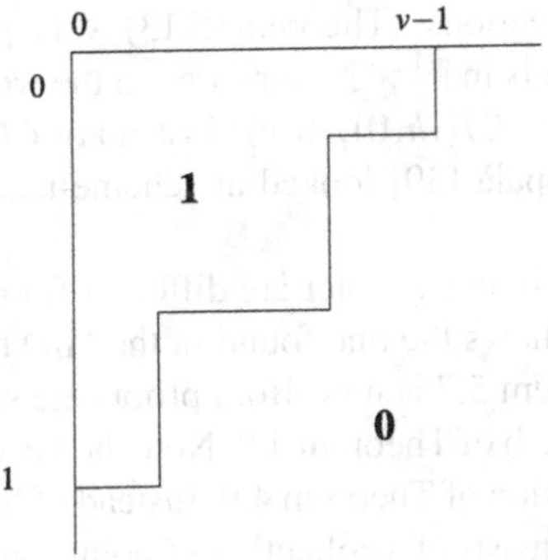

number of ones in the $(i-1)$-th row of the above matrix. Equivalently, set

$$\alpha_i = \sum_{j=0}^{\infty} \Delta H(i-1,j) \text{ for } i = 1,\ldots,h.$$

Now let X be any set of points in $\mathbb{P}^1 \times \mathbb{P}^1$ such that X resembles the partition $(\alpha_1,\alpha_2,\ldots,\alpha_h)$. By construction, X is ACM. Furthermore, by Corollary 5.8 or Corollary 5.10, ΔH_X has α_1 ones in the first row, α_2 ones in the second row, and so on. In other words, $\Delta H_X = \Delta H$. So, H is the Hilbert function of an ACM set of points.

By combining the above result with Theorem 2.29 we have a classification of the Hilbert functions of ACM sets of points in $\mathbb{P}^1 \times \mathbb{P}^1$, thus answering the interpolation problem (Problem 1.2) for this class of points.

Corollary 5.17. *Let $H : \mathbb{N}^2 \to \mathbb{N}$ be a numerical function. Then the following are equivalent:*

(i) H is the Hilbert function of an ACM set of points in $\mathbb{P}^1 \times \mathbb{P}^1$.
(ii) ΔH is the Hilbert function of a bigraded artinian quotient of $k[x_1,y_1]$.
(iii) ΔH satisfies the following conditions:

1. $\Delta H(0,0) = 1$,
2. $\Delta H(i,j) = 0$ *or* 1 *for all* $(i,j) \in \mathbb{N}^2$,
3. $\Delta H(i,j) = 1$ *for only finitely many* $(i,j) \in \mathbb{N}^2$, *and*
4. $\Delta H(i,j) = 0$ *implies that* $\Delta H(k,l) = 0$ *for all* $(k,l) \in \mathbb{N}^2$ *with* $(i,j) \preceq (k,l)$.

5.6 Additional notes

The main results of this chapter (Theorems 5.3, 5.7, and 5.17) were first proved by Giuffrida, Maggioni, and Ragusa [36]. Their results, however, were stated in terms of the first difference function ΔH_X (as noted in Remark 5.5). The classification of complete intersections (Theorem 5.13) was part of the folklore. Complete intersections of points in $\mathbb{P}^1 \times \mathbb{P}^1$ appeared in the work of the first author [43] where they were denoted by $CI((h,0),(0,v))$ instead of $CI(h,v)$. More recently, Giuffrida, Maggioni, and Zappalà [39] looked at scheme-theoretic complete intersections in $\mathbb{P}^1 \times \mathbb{P}^1$.

The proofs used in this chapter are different from those found in [36]. The proof for Theorem 5.3 follows the one found in the PhD thesis of the second author [92]. The proof of Theorem 5.7 is new. Both proofs are similar in that they use the Basic Double Link approach of Theorem 4.9. Note, however, that the proof of Theorem 5.3 uses a modified version of Theorem 4.9. Instead of breaking X into two sets of points Y and Z where Y consists of a collection of points on a single ruling, we are breaking X into two sets of points Y and Z where Y is a complete intersection of points.

There is a natural generalization of Theorem 5.16, as worked out by the second author [94]. Precisely, a multigraded function $H : \mathbb{N}^r \to \mathbb{N}$ is the Hilbert function of an ACM set of points in $\mathbb{P}^{n_1} \times \cdots \times \mathbb{P}^{n_r}$ if and only if the first difference function H, defined by

$$\Delta H(i_1,\ldots,i_r) := \sum_{\underline{0} \preceq \underline{l}=(l_1,\ldots,l_r) \preceq (1,\ldots,1)} (-1)^{|\underline{l}|} H(i_1 - l_1,\ldots,i_r - l_r),$$

where $|\underline{l}| = l_1 + \cdots + l_r$, is the Hilbert function of an artinian quotient of the $\mathbb{N}^k$-graded ring

$$R = k[x_{1,1},\ldots,x_{1,n_1},\ldots,x_{r,1},\ldots,x_{r,n_r}].$$

The proof of this result uses an approach much different from the one we have presented here for $\mathbb{P}^1 \times \mathbb{P}^1$. For the general case, one starts with a monomial ideal in a multigraded artinian quotient, and "lifts" the monomial ideal to a set of points in $\mathbb{P}^{n_1} \times \cdots \times \mathbb{P}^{n_r}$. For the singly graded case, this approach was used by Geramita, Gregory, and Roberts [32] to link the Hilbert functions of points in $\mathbb{P}^n$ to the Hilbert functions of artinian quotients of $k[x_1,\ldots,x_n]$. To generalize this result, one uses the generalized lifting approach developed by Migliore and Nagel [72] to get the desired result. For this chapter, we do not need this technique because we can make use of the Basic Double Link.

While at first glance the above result appears to answer the interpolation problem for ACM sets of points in $\mathbb{P}^{n_1} \times \cdots \times \mathbb{P}^{n_r}$, we do not get an answer because there is no known classification of these artinian quotients in general. Besides the case of the Hilbert functions of the bigraded artinian quotients of $k[x_1,y_1]$ which is given by

Theorem 2.29, we can determine the Hilbert functions of the artinian quotients of the $\mathbb{N}^r$-graded rings $k[x_1,\ldots,x_r]$, and the bigraded artinian quotients of $k[x_1,y_1,\ldots,y_m]$ (see [92] for the details). So, the following problem remains open:

Problem 5.18. *What can be the Hilbert function of an ACM set of points in* $\mathbb{P}^{n_1} \times \cdots \times \mathbb{P}^{n_r}$*?*

Given the difficulty of determining what functions can be the Hilbert function of an ACM sets of points in $\mathbb{P}^{n_1} \times \cdots \times \mathbb{P}^{n_r}$, it is not surprising that there is even less known about the mulitgraded minimal free resolutions of these ideals. In fact, even for non-ACM sets of points X in $\mathbb{P}^1 \times \mathbb{P}^1$ there are very few results about the bigraded minimal free resolutions of the associated ideals $I(X)$. In the papers [37, 38] Giuffrida, Maggioni, and Ragusa worked out the bigraded minimal free resolution of $I(X)$ when X is a set of points in generic position in $\mathbb{P}^1 \times \mathbb{P}^1$. In [11] Bonacini and Marino show that one can start with the bigraded minimal free resolution of $I(X)$ when X is an ACM set of points in $\mathbb{P}^1 \times \mathbb{P}^1$, and by removing points from X under suitable hypotheses to form a set of points Y, one can also determine the bigraded minimal free resolution of $I(Y)$. In general, this new set of points Y is rarely ACM.

Chapter 6
Fat points in $\mathbb{P}^1 \times \mathbb{P}^1$

In the previous chapters we focused on sets of reduced points X in $\mathbb{P}^1 \times \mathbb{P}^1$, that is, $I(X) = \sqrt{I(X)}$. We now relax this condition to study "sets of fat points". Roughly speaking, given a set of reduced points X, we assign to each $P_i \in X$ a positive integer m_i, called its multiplicity, and we consider the ideal $I(Z) = \bigcap_{i=1}^{s} I(P_i)^{m_i}$. We can then ask similar questions for the set of fat point Z defined by $I(Z)$: when is Z arithmetically Cohen-Macaulay? If Z is ACM, what is its Hilbert function H_Z? What is the bigraded minimal free resolution of $I(Z)$? We answer these questions in this chapter.

Like the case of reduced points, we are able to answer these questions directly from an appropriate combinatorial description of the set of points. We begin this chapter with a formal introduction to fat points. We then introduce the combinatorial descriptions that we will exploit. These descriptions generalize the tuples α_X, β_X, and the set S_X introduced in Chapter 3. Our main result, Theorem 6.21, is a classification of ACM sets of fat points which generalizes Theorem 4.11 to fat points. We conclude this chapter with a new proof for the bigraded minimal free resolution of ACM sets of fat points in $\mathbb{P}^1 \times \mathbb{P}^1$ (Theorem 6.27), and we use this result to compute the Hilbert functions of ACM sets of fat points in $\mathbb{P}^1 \times \mathbb{P}^1$ (Theorem 6.28).

6.1 Basics of fat points

We formally introduce fat points, and collect together the results that we will require for the rest of this chapter.

Let X be any set of s points in $\mathbb{P}^1 \times \mathbb{P}^1$. Let $\pi_1 : \mathbb{P}^1 \times \mathbb{P}^1 \to \mathbb{P}^1$ denote the projection morphism defined by $P = A \times B \mapsto A$. Similarly, let π_2 denote the projection morphism $P = A \times B \mapsto B$. The set $\pi_1(X) = \{A_1, \ldots, A_h\}$ is the set of $h \leq s$ distinct first coordinates that appear in X, while $\pi_2(X) = \{B_1, \ldots, B_v\}$ is the set of $v \leq s$

© The Authors 2015
E. Guardo, A. Van Tuyl, *Arithmetically Cohen-Macaulay Sets of Points in* $\mathbb{P}^1 \times \mathbb{P}^1$,
SpringerBriefs in Mathematics, DOI 10.1007/978-3-319-24166-1_6

distinct second coordinates. When $P \in X$, we sometimes write $P = P_{i,j}$ to mean that $P = A_i \times B_j$. As in the previous chapters, we let H_{A_i} denote the degree $(1,0)$ form that vanishes at all the points of X which have first coordinate A_i for $i = 1,\dots,h$. Similarly, for $j = 1,\dots,v$, let V_{B_j} denote the degree $(0,1)$ form that vanishes at all the points whose second coordinate is B_j.

If X is a finite set of s points in $\mathbb{P}^1 \times \mathbb{P}^1$, and $m_{i_1,j_1},\dots,m_{i_s,j_s}$ are s positive integers, then we let Z denote the subscheme of $\mathbb{P}^1 \times \mathbb{P}^1$ defined by the saturated bihomogeneous ideal

$$I(Z) = \bigcap_{P_{i,j} \in X} I(P_{i,j})^{m_{i,j}} = \bigcap_{P_{i,j} \in X} (H_{A_i}, V_{B_j})^{m_{i,j}}.$$

Definition 6.1. With the notation as above, we call Z *a set of fat points* of $\mathbb{P}^1 \times \mathbb{P}^1$. We say Z is a *homogeneous* set of fat points if all the $m_{i,j}$'s are equal. The *support* of Z, written $\mathrm{Supp}(Z)$, is the set of points X.

The set of fat points Z will sometimes be denoted by

$$Z = \{(P_{i,j}; m_{i,j}) \mid P_{i,j} \in X\} = \{(P_{i_1,j_1}; m_{i_1,j_1}), (P_{i_2,j_2}; m_{i_2,j_2}), \dots, (P_{i_s,j_s}; m_{i_s,j_s})\}.$$

Remark 6.2. If $X = \mathrm{Supp}(Z)$, then $I(X) = \sqrt{I(Z)}$. Because $I(Z) \neq \sqrt{I(Z)}$ for a set of fat points, we sometimes refer to Z as a set of nonreduced points.

Many results about reduced sets of points continue to hold for the non-reduced case. We begin with a result on the ideal of a single fat point.

Lemma 6.3. *Let $P \in \mathbb{P}^1 \times \mathbb{P}^1$ be a point and $m \geq 1$ an integer. Then $I(P)^m$ is an $I(P)$-primary ideal.*

Proof. It is enough to show that $I(P)^m$ is primary, since $\sqrt{I(P)^m} = I(P)$. After a change of coordinates, we can assume that $P = [1:0] \times [1:0]$, and thus, $I(P) = (x_1, y_1)$. Because the ideal $I(P)$ is a monomial ideal generated by a subset of the variables, we can use [98, Corollary 5.1.9] to deduce that $I(P)^m$ is primary.

Remark 6.4. For an arbitrary prime ideal P in a ring S, it is not always true that P^m is P-primary for all m; see [62, Example 8.27] for an example. The ideal P^m may have some embedded components.

Corollary 6.5. *Let $Z \subseteq \mathbb{P}^1 \times \mathbb{P}^1$ be a set of fat points. Then the set of minimal associated primes of $I(Z)$ is $\{I(P) \mid P \in \mathrm{Supp}(Z)\}$.*

Proof. By construction, $I(Z) = \bigcap_{i=1}^{s} I(P_i)^{m_i}$. By Corollary 6.3, each ideal $I(P_i)^{m_i}$ is $I(P_i)$-primary for $i = 1,\dots,s$, so this decomposition is an irredundant primary decomposition of $I(Z)$. Moreover, each ideal $I(P_i)$ must be minimal.

Lemma 6.6. *Let $P = A \times B \in \mathbb{P}^1 \times \mathbb{P}^1$ be a point and $m \geq 1$ an integer. If $I(P) = (H_A, V_B)$, then $I(P)^m : (H_A) = I(P)^{m-1}$ (where we take $I(P)^0 = R$).*

Proof. Let $F \in I(P)^{m-1}$. Because $H_A \in I(P)$, $FH_A \in I(P)^m$ and hence $F \in I(P)^m : (H_A)$.

For the reverse inclusion, we first make a change of coordinates so that $P = [1:0] \times [1:0]$, and thus $I(P) = (x_1, y_1)$ and $H_A = x_1$. Suppose that $F \in I(P)^m : (H_A) = (x_1, y_1)^m : (x_1)$. Because $(x_1, y_1)^m : (x_1)$ is a monomial ideal, we can take F to be a monomial, i.e., $F = x_0^{a_0} x_1^{a_1} y_0^{b_0} y_1^{b_1}$ for some non-negative integers a_0, a_1, b_0, and b_1. Because $Fx_1 = x_0^{a_0} x_1^{a_1+1} y_0^{b_0} y_1^{b_1} \in (x_1, y_1)^m$, we have $a_1 + 1 + b_1 \geq m$, and thus $a_1 + b_1 \geq m - 1$. But this inequality then implies that $F \in (x_1, y_1)^{m-1}$, as desired.

Lemma 6.7. *Let $Z \subseteq \mathbb{P}^1 \times \mathbb{P}^1$ be a set of fat points. Then there exists a form $L \in R_{1,0}$ such that $\overline{L}$ is a nonzero-divisor in $R/I(Z)$.*

Proof. Let $Y = \pi_1(X) = \{A_1, \ldots, A_h\}$ be the distinct first coordinates that appear in $X = \text{Supp}(Z)$, and view Y as a set of points in $\mathbb{P}^1$. Let L be any linear form of $k[x_0, x_1]$ that does not vanish at any of the points in Y. Because k is infinite, such a linear form exists. We now view L as an element of $R = k[x_0, x_1, y_0, y_1]$, and thus $L \in R_{1,0}$.

Suppose that $FL \in I(Z)$. For any $P \in X$, we have $FL \in I(P)^m$ where m is the multiplicity associated with P. Now $P = A \times B$ with $A \in Y$. Furthermore, $L(P) = L(A)$ because L is a polynomial only in the variables $\{x_0, x_1\}$. By our choice of L, $L(A) \neq 0$. Thus, $L^k \notin I(P)^m$ for all integers $k \geq 1$. Since $I(P)^m$ is primary by Lemma 6.3, $F \in I(P)^m$. Because this is true for all $P \in \text{Supp}(Z)$, $F \in I(Z)$. So $\overline{L}$ is not a zero-divisor in $R/I(Z)$.

Remark 6.8. A similar argument can be used to show that there exists a form $L' \in R_{0,1}$ such that $\overline{L'}$ is a nonzero-divisor in $R/I(Z)$.

Like reduced points, there are only two possible values for the depth of $R/I(Z)$.

Lemma 6.9. *Let $Z \subseteq \mathbb{P}^1 \times \mathbb{P}^1$ be a set of fat points. Then*

$$1 \leq \text{depth}(R/I(Z)) \leq \text{K-dim}(R/I(Z)) = 2.$$

Proof. By Lemma 6.7, there is a nonzero-divisor on $R/I(Z)$, which implies that $1 \leq \text{depth}(R/I(Z))$. To complete the proof, we need to show $\text{K-dim}(R/I(Z)) = 2$.

By Corollary 6.5, the set of minimal associated primes are

$$\text{Min}(R/I(Z)) = \{I(P) \mid P \in \text{Supp}(Z)\}.$$

Thus, we have

$$\begin{aligned} \text{K-dim}(R/I(Z)) &= \sup\{\text{K-dim}(R/I(P)) \mid P \in \text{Min}(R/I(Z))\} \\ &= \sup\{\text{K-dim}(R/I(P)) \mid P \in \text{Supp}(Z)\} = 2 \end{aligned}$$

because $\text{K-dim}(R/I(P)) = 2$ for all points $P \in \text{Supp}(Z)$.

Lemma 6.7 can be used to prove the following basic results about H_Z, the Hilbert function of $R/I(Z)$. The proofs are almost word-for-word the same as Theorem 3.27.

Theorem 6.10. *Let $Z \subseteq \mathbb{P}^1 \times \mathbb{P}^1$ be a set of fat points with Hilbert function H_Z.*

(i) $H_Z(i,j) \leq H_Z(i+1,j)$ for all $(i,j) \in \mathbb{N}^2$.
(ii) $H_Z(i,j) \leq H_Z(i,j+1)$ for all $(i,j) \in \mathbb{N}^2$.
(iii) If $H_Z(i,j) = H_Z(i+1,j)$, then $H_Z(i,j) = H_Z(i+a,j)$ for all $a \in \mathbb{N}$.
(iv) If $H_Z(i,j) = H_Z(i,j+1)$, then $H_Z(i,j) = H_Z(i,j+b)$ for all $b \in \mathbb{N}$.

Proof. It will suffice to prove statements (i) and (iii), because the other statements are proved similarly, but use the other grading.

(i) By Lemma 6.7 there exists a form $L \in R_{1,0}$ such that $\overline{L}$ is a nonzero-divisor of $R/I(Z)$. Because $\deg L = (1,0)$, we have the following map between vector spaces

$$\times L : (R/I(Z))_{i,j} \xrightarrow{\times \overline{L}} (R/I(Z))_{i+1,j}$$

that is injective for all $(i,j) \in \mathbb{N}^2$. Because the map is injective, we must have $H_Z(i,j) \leq H_Z(i,j+1)$.

(iii) Again, let $\overline{L}$ be the nonzero-divisor of $R/I(Z)$ given by Lemma 6.7. For each $(i,j) \in \mathbb{N}^2$, we have the following short exact sequence of vector spaces:

$$0 \longrightarrow (R/I(Z))_{i,j} \xrightarrow{\times \overline{L}} (R/I(Z))_{i+1,j} \longrightarrow (R/(I(Z),L))_{i+1,j} \longrightarrow 0.$$

If $H_Z(i,j) = H_Z(i+1,j)$, then the multiplication map $\times \overline{L}$ is an isomorphism of vector spaces, and thus, $(R/(I(Z),L))_{i+1,j} = 0$. But then for any $a \geq 1$, we have $(R/(I(Z),L))_{i+a,j} = 0$ as well. Hence, from the short exact sequence

$$0 \longrightarrow (R/I(Z))_{i+a-1,j} \xrightarrow{\times \overline{L}} (R/I(Z))_{i+a,j} \longrightarrow (R/(I(Z),L))_{i+a,j} \longrightarrow 0$$

we deduce that $H_Z(i,j) = \dim_k(R/I(Z))_{i,j} = \dim_k(R/I(Z))_{i+a,j} = H_Z(i+a,j)$ for all $a \geq 1$.

6.2 Two combinatorial descriptions of fat points

We introduce two combinatorial descriptions one can attach to a set of fat points in $\mathbb{P}^1 \times \mathbb{P}^1$. These descriptions will generalize the definitions of the tuples α_X and β_X, and the set S_X introduced in Chapter 3. Before introducing the notation, we introduce a running example for this section which will be used to illustrate our constructions.

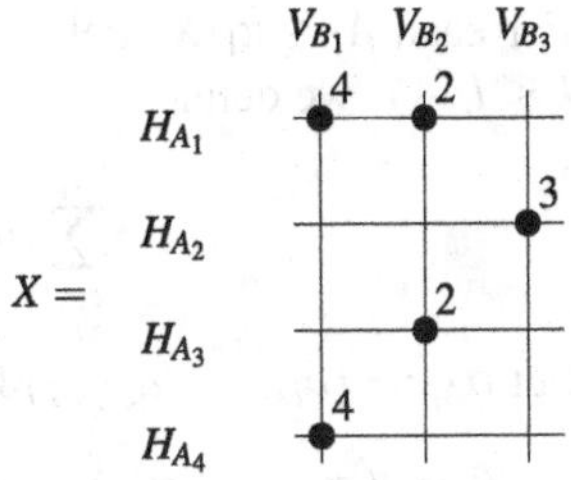

Fig. 6.1 Running example of fat points

Example 6.11. Let $\{A_1, A_2, A_3, A_4\}$ be four distinct points of $\mathbb{P}^1$ and let $\{B_1, B_2, B_3\}$ be three distinct points of $\mathbb{P}^1$. We now consider the set of fat points Z defined by the following ideal

$$I(Z) = (H_{A_1}, V_{B_1})^4 \cap (H_{A_1}, V_{B_2})^2 \cap (H_{A_2}, V_{B_3})^3 \cap (H_{A_3}, V_{B_2})^2 \cap (H_{A_4}, V_{B_1})^4.$$

Alternatively, we can represent this set of fat points by

$$Z = \{(P_{1,1};4), (P_{1,2};2), (P_{2,3};3), (P_{3,2};2), (P_{4,1};4)\} \text{ with } P_{i,j} = A_i \times B_j.$$

As in previous chapters, we can "draw" sets of fat points. In particular, we draw the support of Z using Convention 3.9, and then we label each point $P \in \text{Supp}(Z)$ with its corresponding multiplicity. Thus, we can visualize our running example Z as in Figure 6.1.

We now turn to our generalizations of the tuples α_X and β_X. Although at first glance the generalized definition may look difficult, what we are doing is easy to summarize. For each horizontal ruling, sum up all the multiplicities of the points on that ruling. Then subtract one from all the multiplicities, and add up the resulting numbers. Subtract one again from all the new multiplicities (provided the number is positive), and again add up the resulting numbers. Continue until all the multiplicities have been reduced to zero. We then do the same for the vertical rulings. The definition below simply formalizes this procedure. In what follows, we use that notation

$$(n)_+ := \max\{0, n\}.$$

Let $Z \subseteq \mathbb{P}^1 \times \mathbb{P}^1$ be a set of fat points with support $X = \text{Supp}(Z)$, and suppose that $\pi_1(X) = \{A_1, \ldots, A_h\}$. For each $A_i \in \pi_1(X)$, define

$$Z_{1,A_i} := \{(P_{i,j_1}; m_{i,j_1}), (P_{i,j_2}; m_{i,j_2}), \ldots, (P_{i,j_{v_i}}; m_{i,j_{v_i}})\}$$

where $P_{i,j_k} = A_i \times B_{j_k}$ are those points of $\text{Supp}(Z)$ whose first coordinate is A_i. Thus $\pi_1(\text{Supp}(Z_{1,A_i})) = \{A_i\}$, and furthermore it follows that

$$I(Z) = \bigcap_{i=1}^{h} I_{Z_{1,A_i}}.$$

For each $A_i \in \pi_1(X)$ define $l_i := \max\{m_{i,j_1}, \ldots, m_{i,j_{v_i}}\}$. Then, for each integer $0 \leq k \leq l_i - 1$, we define

$$a_{i,k} := \sum_{e=1}^{v_i} (m_{i,j_e} - k)_+ \qquad \text{where } (n)_+ := \max\{n, 0\}.$$

Set $\alpha_{A_i} := (a_{i,0}, \ldots, a_{i,l_i-1})$ for each $A_i \in \pi_1(X)$. Define

$$\tilde{\alpha}_Z := (\alpha_{A_1}, \ldots, \alpha_{A_h}) = (a_{1,0}, \ldots, a_{1,l_1-1}, a_{2,0}, \ldots, a_{2,l_2-1}, \ldots, a_{h,0}, \ldots, a_{h,l_h-1}).$$

Finally, we define α_Z to be the $(l_1 + \cdots + l_h)$-tuple one gets by rearranging the elements of $\tilde{\alpha}_Z$ in non-increasing order.

The tuple β_Z is defined similarly, but is defined in terms of the vertical rulings instead of the horizontal rulings. For completeness, we include the details. For each $B_j \in \pi_2(X)$, set

$$Z_{2,B_j} := \{(P_{i_1,j}; m_{i_1,j}), (P_{i_2,j}; m_{i_2,j}), \ldots, (P_{i_{h_j},j}; m_{i_{h_j},j})\}$$

where $P_{i_k,j} = A_{i_k} \times B_j$ are those points of Supp(Z) whose second coordinate is B_j. Thus $\pi_2(\mathrm{Supp}(Z_{2,B_j})) = \{B_j\}$. For each $B_j \in \pi_2(X)$ set $l'_j = \max\{m_{i_1,j}, \ldots, m_{i_{h_j},j}\}$. Then, for each integer $0 \leq k \leq l'_j - 1$, we define

$$b_{j,k} := \sum_{e=1}^{h_j} (m_{i_e,j} - k)_+ \qquad \text{where } (n)_+ := \max\{n, 0\}.$$

Let $\beta_{B_j} := (b_{j,0}, \ldots, b_{j,l'_j-1})$ for each $B_j \in \pi_2(X)$. Define

$$\tilde{\beta}_Z := (\beta_{B_1}, \ldots, \beta_{B_v}) = (b_{1,0}, \ldots, b_{1,l'_1-1}, b_{2,0}, \ldots, b_{2,l'_2-1}, \ldots, b_{v,0}, \ldots, b_{v,l'_v-1}).$$

We then define β_Z to be the $(l'_1 + \cdots + l'_h)$-tuple one gets by rearranging the elements of $\tilde{\beta}_Z$ in descending order.

Remark 6.12. When all the $m_{i,j} = 1$, i.e., Z is a set of reduced points, then α_Z and β_Z will agree with the definition of α_X and β_X from Chapter 3.

Example 6.13. We will calculate α_Z and β_Z for the set of fat points of Example 6.11. Note that $\pi_1(\mathrm{Supp}(Z)) = \{A_1, A_2, A_3, A_4\}$. Then

$$Z_{1,A_1} = \{(P_{11}; 4), (P_{12}; 2)\}.$$

We set $l_1 := \max\{4, 2\} = 4$. Then

$$\begin{aligned} a_{1,0} &= 4 + 2 = 6 \\ a_{1,1} &= (4-1)_+ + (2-1)_+ = 4 \\ a_{1,2} &= (4-2)_+ + (2-2)_+ = 2 \\ a_{1,3} &= (4-3)_+ + (2-3)_+ = 1. \end{aligned}$$

Hence, $\alpha_{A_1} = (6,4,2,1)$. Note that what we have done here is sum up the multiplicities on the $(1,0)$-line defined by H_{A_1}. We then subtracted one from each multiplicity, and then added up the resulting new multiplicities. We continue this procedure until each multiplicity has been reduced to zero.

For A_2, A_3, and A_4, we get $\alpha_{A_2} = (3,2,1)$, $\alpha_{A_3} = (2,1)$, $\alpha_{A_4} = (4,3,2,1)$. Hence

$$\tilde{\alpha}_Z = (6,4,2,1,3,2,1,2,1,4,3,2,1),$$

and thus,

$$\alpha_Z = (6,4,4,3,3,2,2,2,2,1,1,1,1).$$

For $B_1, B_2, B_3 \in \pi_2(\text{Supp}(X))$, we will find $l'_1 = 4$, $l'_2 = 2$ and $l'_3 = 3$, respectively. So, we have $\beta_{B_1} = (8,6,4,2)$, $\beta_{B_2} = (4,2)$, and $\beta_{B_3} = (3,2,1)$, and therefore,

$$\tilde{\beta}_Z = (8,6,4,2,4,2,3,2,1)$$

and thus,

$$\beta_Z = (8,6,4,4,3,2,2,2,1).$$

We now generalize S_X to sets of fat points in $\mathbb{P}^1 \times \mathbb{P}^1$. Let Z be a set of fat points whose support is contained in h horizontal rulings, and v vertical rulings. If $\pi_1(\text{Supp}(Z)) = \{A_1, \ldots, A_h\}$ and $\pi_2(\text{Supp}(Z)) = \{B_1, \ldots, B_v\}$, then we can write Z as

$$Z = \{(P_{i,j}; m_{i,j}) \mid 1 \leq i \leq h,\ 1 \leq j \leq v\}$$

where $P_{i,j} = A_i \times B_j$ and $m_{i,j} > 0$ if and only if $P_{i,j} \in \text{Supp}(Z)$ (we will write $m_{i,j} = 0$ to denote the fact that $P_{i,j}$ does not appear in the support).

For each $\ell \in \mathbb{N}$, and for each tuple (i,j) with $1 \leq i \leq h$ and $1 \leq j \leq v$, we define

$$t_{i,j}(\ell) := (m_{i,j} - \ell)_+ = \max\{0, m_{i,j} - \ell\}.$$

The set S_Z is then defined to be the set of nonzero v-tuples

$$S_Z = \{(t_{i,1}(\ell), \ldots, t_{i,v}(\ell)) \mid 1 \leq i \leq h,\ \ell \in \mathbb{N}\}.$$

Again, while the definition may seem complicated, we can easily describe our construction. For each horizontal rule H_A that intersects the support of Z, construct the v-tuple whose entries consist of the multiplicities of the points $A \times B_1, \ldots, A \times B_v$. Then subtract one from each positive element of the tuple to create a new v-tuple. Continue subtracting one from each positive element to create new tuples until all entries are zero. The set S_Z is the collection of these nonzero v-tuples.

Example 6.14. Returning to the running example of Example 6.11, we compute S_Z. Since three vertical lines contain the support of Z, the set S_Z will consist of 3-tuples.

Consider the fat points $\{(P_{1,1};4),(P_{1,2};2),(P_{1,3};0)\}$ on the horizontal ruling H_{A_1}. Then we have

$$\begin{array}{lll} t_{1,1}(0) = (4-0)_+ = 4 & t_{1,2}(0) = (2-0)_+ = 2 & t_{1,3}(0) = (0-0)_+ = 0 \\ t_{1,1}(1) = (4-1)_+ = 3 & t_{1,2}(0) = (2-1)_+ = 1 & t_{1,3}(0) = (0-1)_+ = 0 \\ t_{1,1}(2) = (4-2)_+ = 2 & t_{1,2}(0) = (2-2)_+ = 0 & t_{1,3}(0) = (0-2)_+ = 0 \\ t_{1,1}(3) = (4-3)_+ = 1 & t_{1,2}(0) = (2-3)_+ = 0 & t_{1,3}(0) = (0-3)_+ = 0 \end{array}$$

Furthermore, for all $\ell \geq 4$, $t_{1,j}(\ell) = 0$ for all $j = 1,2,3$. So, the 3-tuples $(4,2,0)$, $(3,1,0)$, $(2,0,0)$, and $(1,0,0)$ all belong to S_Z.

We repeat this procedure for each horizontal ruling that meets Supp(Z). For example, the 3-tuples that we construct from the second horizontal ruling are

$$(0,0,3),(0,0,2),(0,0,1).$$

Continuing in this fashion, we fill find that

$$\begin{aligned} S_Z = \{&(4,2,0),(3,1,0),(2,0,0),(1,0,0),(0,0,3),(0,0,2),(0,0,1), \\ &(0,2,0),(0,1,0),(4,0,0),(3,0,0),(2,0,0),(1,0,0)\}. \end{aligned}$$

As in Chapter 3, our two combinatorial descriptions are linked. Returning to the above example, note that if you sum up the entries of each tuple in S_Z, and rearrange in non-increasing order, you will construct α_Z. In fact, a more general version of Theorem 3.21 holds. We first require a combinatorial lemma.

Lemma 6.15. *Let* $\alpha = (\alpha_1,\ldots,\alpha_n)$, $\beta = (\beta_1,\ldots,\beta_m)$, *and suppose that* $\alpha,\beta \vdash s$. *If* $\alpha^* = \beta$, *then*

(*i*) $\alpha_1 = |\beta|$.
(*ii*) $\beta_1 = |\alpha|$.
(*iii*) *if* $\alpha' = (\alpha_2,\ldots,\alpha_n)$ *and* $\beta' = (\beta_1 - 1,\ldots,\beta_{\alpha_2} - 1)$, *then* $(\alpha')^* = \beta'$.

Proof. The proof of (i) and (ii) are the same. We do (ii). By definition, $\alpha_1^* = |\{\alpha_i \in \alpha \mid \alpha_i \geq 1\}|$. Because $\alpha \vdash s$, $\alpha_i \geq 1$ for every i, and so $\alpha_1^* = n = |\alpha|$. But $\alpha^* = \beta$ implies $\alpha_1^* = \beta_1$, thus completing the proof.

For (iii), because $\alpha_j^* = \beta_j$, we have $|\{\alpha_i \in \alpha \mid \alpha_i \geq j\}| = \beta_j$ for every $1 \leq j \leq \alpha_1$. Now $\alpha_1 \geq \alpha_i$ for every coordinate α_i of $\alpha' = (\alpha_2,\ldots,\alpha_n)$. Thus, we can rewrite α_j^* as

$$\alpha_j^* = \begin{cases} |\{\alpha_i \in \alpha' \mid \alpha_i \geq j\}| + 1 = (\alpha')_j^* + 1 & \text{if } 1 \leq j \leq \alpha_2 \\ 1 & \text{if } \alpha_2 < j \leq \alpha_1 \end{cases}.$$

Hence

$$\beta_j - 1 = \alpha_j^* - 1 = \begin{cases} (\alpha')_j^* + 0 & \text{if } 1 \leq j \leq \alpha_2 \\ 0 & \text{if } \alpha_2 < j \leq \alpha_1 \end{cases}.$$

The conclusion now follows.

Theorem 6.16. *Let $Z \subseteq \mathbb{P}^1 \times \mathbb{P}^1$ be a set of fat points, and let α_Z, β_Z, and S_Z be constructed from Z as described above. Then the following are equivalent:*

(i) $\alpha_Z^* = \beta_Z$ *(or* $\alpha_Z = \beta_Z^*$*)*
(ii) S_Z *has no incomparable elements with respect to the partial order* $\succeq$ *on* $\mathbb{N}^v$.

Proof. (i) $\Rightarrow$ (ii) Suppose that $Z = \{(P_{i,j}; m_{i,j}) \mid 1 \leq i \leq h, 1 \leq j \leq v\}$ where $m_{i,j}$ are non-negative numbers and $\alpha_Z^* = \beta_Z$. (Note that we are including the points $(P_{i,j};0)$ where $m_{i,j} = 0$ means that the point $P_{i,j}$ does not appear in the support.) We will work by induction on $\beta_1 = \max\left\{\sum_{i=1}^h m_{i,j}\right\}_{j=1}^v$.

If $\beta_1 = 1$, then Z is a set of v distinct reduced points and $\beta_Z = (\underbrace{1,\ldots,1}_{v})$. Because $\alpha_Z^* = \beta_Z$, we have $\alpha_Z = (v)$. Thus $Z = \{A \times B_1, \ldots, A \times B_v\}$, in which case we see that $S_Z = \{(1,\ldots,1)\}$, and that this set has no incomparable elements.

Now suppose that $\beta_1 > 1$ and that the theorem holds for all sets of fat points Y with $\alpha_Y^* = \beta_Y$, and the first coordinate of β_Y is less than β_1.

Let k be the index in $\{1,\ldots,h\}$ such that $\alpha_1 = \sum_{j=1}^v m_{k,j}$.

Claim. $m_{k,j} = \max\{m_{1,j},\ldots,m_{h,j}\} > 0$ for each $j = 1,\ldots,v$.

Proof of the Claim. Set $l'_j = \max\{m_{1,j},\ldots,m_{h,j}\}$ for $j = 1,\ldots,v$. From the construction of β_Z, we have $|\beta_Z| = l'_1 + \cdots + l'_v$. Since $\alpha_Z^* = \beta_Z$, by Lemma 6.15 $\alpha_1 = l'_1 + \cdots + l'_v$. Now suppose that $l'_c > m_{k,c}$ for some $c \in \{1,\ldots,v\}$. Because $l'_j \geq m_{k,j}$ for each $j = 1,\ldots,v$, we would then have

$$\alpha_1 = l'_1 + \cdots + l'_c + \cdots + l'_v > m_{k,1} + \cdots + m_{k,c} + \cdots + m_{k,v} = \alpha_1$$

Because of this contradiction, the claim holds. □

Let $Y = \{(P_{i,j}; m'_{i,j}) \mid 1 \leq i \leq h, 1 \leq j \leq v\}$ be a set of fat points where

$$m'_{i,j} = \begin{cases} m_{i,j} & i \neq k \\ m_{k,j} - 1 & i = k \end{cases}.$$

By the claim $m_{k,j} - 1 \geq 0$ for all $j = 1,\ldots,v$. Let β be the first coordinate of β_Y. Then $\beta < \beta_1$. In fact, for each $j = 1,\ldots,v$, we have

$$\sum_{i=1}^h m'_{i,j} = m'_{k,j} + \sum_{i \neq k} m_{i,j} = \left(\sum_{i=1}^h m_{i,j}\right) - 1.$$

Furthermore, if $\alpha_Z = (\alpha_1,\ldots,\alpha_m)$ and $\beta_Z = (\beta_1,\ldots,\beta_{m'})$, then from our construction Y we have $\alpha_Y = (\alpha_2,\ldots,\alpha_m)$ and $\beta_Y = (\beta_1 - 1,\ldots,\beta_{\alpha_2} - 1)$. By Lemma 6.15, $\alpha_Y^* = \beta_Y$, and so by induction S_Y has no incomparable elements with respect to $\succeq$, or equivalently, S_Y is totally ordered with respect to $\succeq$.

The set S_Z is now obtained from S_Y by adding the tuple $(m_{k,1},\ldots,m_{k,v})$. It then follows from our claim that this element is larger than every other element of S_Y with respect to the ordering $\succeq$. So S_Z is totally ordered, as desired.

(ii) $\Rightarrow$ *(i)*. We do induction on $|S_Z|$. If $|S_Z| = 1$, then the only possibility is $S_Z = \{(\underbrace{1,\ldots,1}_{v})\}$. Indeed, if $m_{i,j} > 1$, then S_Z would contain at least two elements. Similarly, if at least two horizontal rulings meet Z, then S_Z would contain at least two elements. It then follows that $Z = \{A \times B_1, \ldots, A \times B_v\}$, from which we deduce that $\alpha_Z = (v)$ and $\beta_Z = (\underbrace{1,\ldots,1}_{v})$, and thus, $\alpha_Z^* = \beta_Z$.

Suppose that $|S_Z| > 1$. Because S_Z has no incomparable elements with respect to $\succeq$, it has a largest element, say $(m_{k,1},\ldots,m_{k,v})$. Now consider the set of fat points Y given by $Y = \{(P_{i,j}; m'_{i,j}) \mid 1 \le i \le h, 1 \le j \le v\}$ where

$$m'_{i,j} = \begin{cases} m_{i,j} & i \neq k \\ m_{k,j} - 1 & i = k \end{cases}.$$

By our construction $S_Y = S_Z \setminus \{(m_{k,1},\ldots,m_{k,v})\}$, and thus S_Y will also have no incomparable elements with respect to $\succeq$. Thus, by induction $\alpha_Y^* = \beta_Y$.

We now compute α_Z and α_Z^*. Now $\alpha_Z = (\alpha_1,\ldots,\alpha_l)$ where $l = l_1 + \cdots + l_h$ and $l_i = \max\{m_{i,1},\ldots,m_{i,v}\}$. Since $(m_{k,1},\ldots,m_{k,v})$ is the maximal element of S_Z, and because S_Z has no incomparable elements, $\alpha_1 = m_{k,1} + \cdots + m_{k,v}$. In addition, because $(m_{k,1},\ldots,m_{k,v})$ is the maximal element of S_Z, we have $l'_j = m_{k,j} = \max\{m_{1,j},\ldots,m_{h,j}\}$ for each $j = 1,\ldots,v$. So $|\beta_Z| = m_{k,1} + \cdots + m_{k,v}$. It then follows from Lemma 6.15 that $\alpha_Z = (|\beta_Z|, \alpha_2, \ldots, \alpha_l)$. Moreover, from our construction of Y, $\alpha_Y = (\alpha_2,\ldots,\alpha_l)$. We can thus write α_j^* of α_Z^* as

$$\alpha_j^* = \begin{cases} |\{\alpha_i \in \alpha_Y \mid \alpha_i \ge j\}| + 1 = (\alpha_Y)_j^* + 1 & \text{if } 1 \le j \le \alpha_2 \\ 1 & \text{if } \alpha_2 < j \le |\beta_Z| \end{cases}.$$

Because $\beta_Y = \alpha_Y^*$, we have

$$\alpha_Z^* = ((\beta_Y)_1 + 1, \ldots, (\beta_Y)_{\alpha_2} + 1, \underbrace{1,\ldots,1}_{|\beta_Z| - \alpha_2)}).$$

On the other hand, $\beta_Z = (\beta_1,\ldots,\beta_{l'})$ where $l' = l'_1 + \cdots + l'_v = \alpha_1$. When we compute β_Y for Y we have

$$\beta_Y = (\beta_1 - 1, \ldots, \beta_{\alpha_2} - 1, \beta_{\alpha_2+1} - 1, \ldots, \beta_{\alpha_1} - 1).$$

On the other hand, $|\beta_Y| = \alpha_2$. To see this, let $(m_{g,1},\ldots,m_{g,v})$ be the second largest element of S_Z with respect to $\succeq$. Then $\alpha_2 = m_{g,1} + \cdots + m_{g,v}$. Because $(m_{g,1},\ldots,m_{g,v})$ is the largest element of S_Y, our construction of Y and the definition of β_Y implies $|\beta_Y| = \alpha_2$. So $\beta_{\alpha_2+1} - 1 = \cdots = \beta_{\alpha_1} - 1 = 0$. (Note, we only need to check this when $\alpha_2 < \alpha_1$; if $\alpha_2 = \alpha_1$, we can move to the next step.) But then

$$\begin{aligned}\alpha_Z^* &= ((\beta_Y)_1 + 1, \ldots, (\beta_Y)_{\alpha_2} + 1, \underbrace{1, \ldots, 1}_{|\beta_Z| - \alpha_2}) \\ &= (\beta_1 - 1 + 1, \ldots, \beta_{\alpha_2} - 1 + 1, \beta_{\alpha_2+1} - 1 + 1, \ldots, \beta_{\alpha_1} - 1 + 1) = \beta_Z\end{aligned}$$

which is our desired conclusion.

6.3 Hilbert functions of fat points

We now generalize the results of Section 3.3 by showing that all, but possibly a finite number of values, of the Hilbert function of $R/I(Z)$ when Z is a set of fat points in $\mathbb{P}^1 \times \mathbb{P}^1$ can be computed directly from the tuples α_Z and β_Z. We use the same basic strategy as in Theorem 3.29 to prove the main result of this section.

We begin by computing the Hilbert function of a set of points whose support is contained on a single horizontal ruling.

Theorem 6.17. *Let $Z = \{(P_{1,1}; m_{1,1}), (P_{1,2}; m_{1,2}), \ldots, (P_{1,v}, m_{1,v})\} \subseteq \mathbb{P}^1 \times \mathbb{P}^1$ be a set of fat points whose support is on a single horizontal ruling. If $\alpha_Z = (a_1, \ldots, a_m)$ with $m = \max\{m_{1,j}\}_{j=1}^{v}$ is the tuple associated with Z as above, then the Hilbert function of Z is*

$$H_Z = \begin{bmatrix} 1 & 2 & \cdots & a_1 - 1 & a_1 & a_1 & \cdots \\ 1 & 2 & \cdots & a_1 - 1 & a_1 & a_1 & \cdots \\ \vdots & \vdots & & \vdots & \vdots & \vdots & \ddots \end{bmatrix} + \begin{bmatrix} 0 & 0 & \cdots & 0 & 0 & 0 & \cdots \\ 1 & 2 & \cdots & a_2 - 1 & a_2 & a_2 & \cdots \\ 1 & 2 & \cdots & a_2 - 1 & a_2 & a_2 & \cdots \\ \vdots & \vdots & & \vdots & \vdots & \vdots & \ddots \end{bmatrix}$$

$$+ \cdots + \begin{bmatrix} 0 & 0 & \cdots & 0 & 0 & 0 & \cdots \\ \vdots & \vdots & & \vdots & \vdots & \vdots & \\ 0 & 0 & \cdots & 0 & 0 & 0 & \cdots \\ 1 & 2 & \cdots & a_m - 1 & a_m & a_m & \cdots \\ 1 & 2 & \cdots & a_m - 1 & a_m & a_m & \cdots \\ \vdots & \vdots & & \vdots & \vdots & \vdots & \ddots \end{bmatrix}.$$

Proof. For each $j = 1, \ldots, v$, the ideal associated with $P_{1,j}$ is $I(P_{1,j}) = (H_{A_1}, V_{B_j})$. Set $H = H_{A_1}$ and note that H defines the $(1,0)$ line in $\mathbb{P}^1 \times \mathbb{P}^1$ on which all the points lie. Now for each $0 \leq \ell \leq m - 1$ we set

$$Z_\ell = \{(P_{1,1}; (m_{1,1} - \ell)_+), \ldots, (P_{1,v}; (m_{1,v} - \ell)_+)\}$$

and let $I(Z_\ell)$ be the associated ideal. Thus $Z_0 = Z$. Furthermore, we have the identity

$$H^\ell \cap I(Z) = H^\ell \cdot I(Z_\ell) \quad \text{for each } \ell = 0,\ldots,m-1.$$

Because $H^m \in I(Z)$, we have $0 = \overline{H}^m \cdot S \subseteq \overline{H}^{m-1} \cdot S \subseteq \cdots \subseteq \overline{H} \cdot S \subseteq S$ where $S = R/I(Z)$ and $\overline{H}^i$ denotes the image of H^i in S. It then follows that

$$H_Z(i,j) = \dim_k S_{i,j} = \sum_{\ell=0}^{m-1} \dim_k \left(\frac{\overline{H}^\ell \cdot S}{\overline{H}^{\ell+1} \cdot S} \right)_{i,j}.$$

Now for each $\ell = 0,\ldots,m-1$,

$$\frac{\overline{H}^\ell \cdot S}{\overline{H}^{\ell+1} \cdot S} \cong \frac{H^\ell R}{H^{\ell+1} + H^\ell \cap I(Z)} \cong \frac{H^\ell R}{H^{\ell+1} + H^\ell I(Z_\ell)} \cong \overline{H}^\ell \left(\frac{R}{H + I(Z_\ell)} \right).$$

Hence $\dim_k \left(\frac{\overline{H}^\ell \cdot S}{\overline{H}^{\ell+1} \cdot S} \right)_{i,j} = \dim_k (R/(H+I(Z_\ell)))_{i-\ell,j}$, and thus

$$H_Z(i,j) = \sum_{\ell=0}^{m-1} \dim_k (R/(H+I(Z_\ell)))_{i-\ell,j}.$$

To compute H_Z, we thus need to compute the Hilbert function of $R/(H+I(Z_\ell))$ for each ℓ. We now note that for each ℓ,

$$(H+I(Z_\ell)) = (H, V_{B_1}^{(m_{1,1}-\ell)_+} \cdots V_{B_v}^{(m_{1,v}-\ell)_+}),$$

that is, $(H+I(Z_\ell))$ is a complete intersection generated by forms of degree $(1,0)$ and $(0,a_\ell)$ because $a_\ell = (m_{1,1}-\ell)_+ + \cdots + (m_{1,v}-\ell)_+$. The resolution of $(H+I(Z_\ell))$ is given by Lemma 2.26

$$0 \longrightarrow R(-1,-a_\ell) \longrightarrow R(-1,0) \oplus R(0,-a_\ell) \longrightarrow (H+I(Z_\ell)) \longrightarrow 0.$$

Hence, the Hilbert function of $R/(H+I(Z_\ell)))$ is

$$H_{R/(H+I(Z_\ell))} = \begin{bmatrix} 1 & 2 & \cdots & a_\ell - 1 & a_\ell & a_\ell & \cdots \\ 1 & 2 & \cdots & a_\ell - 1 & a_\ell & a_\ell & \cdots \\ \vdots & \vdots & & \vdots & \vdots & \vdots & \ddots \end{bmatrix}.$$

This now completes the proof.

Theorem 6.18. *Let $Z \subseteq \mathbb{P}^1 \times \mathbb{P}^1$ be a set of fat points with associated tuples $\alpha_Z = (\alpha_1,\ldots,\alpha_l)$ and $\beta_Z = (\beta_1,\ldots,\beta_{l'})$. (Recall that $l = l_1 + \ldots + l_h$ where $l_i = \max\{m_{i,j_1},\ldots,m_{i,j_{v_i}}\}$ and $l' = l'_1 + \ldots + l'_v$ where $l'_j = \max\{m_{i_1,j},\ldots,m_{i_{h_j},j}\}$.)*

(*i*) *For all* $j \in \mathbb{N}$, *if* $i \geq l-1$, *then*

$$H_Z(i,j) = \alpha_1^* + \alpha_2^* + \cdots + \alpha_{j+1}^*$$

where $\alpha_Z^* = (\alpha_1^*, \ldots, \alpha_{\alpha_1}^*)$ *is the conjugate of* α_Z, *and where we make the convention that* $\alpha_t^* = 0$ *if* $t > \alpha_1$.

(*ii*) *For all* $i \in \mathbb{N}$, *if* $j \geq l' - 1$, *then*

$$H_Z(i,j) = \beta_1^* + \beta_2^* + \cdots + \beta_{i+1}^*$$

where $\beta_Z^* = (\beta_1^*, \ldots, \beta_{\beta_1}^*)$ *is the conjugate of* β_Z, *and where we make the convention that* $\beta_t^* = 0$ *if* $t > \beta_1$.

Proof. We will only prove (*i*) because the proof of statement of (*ii*) is similar. Let Z be a set of fat points in $\mathbb{P}^1 \times \mathbb{P}^1$, and let $X = \text{Supp}(Z)$. Our proof is by induction on $h = |\pi_1(X)|$. If $h = 1$, i.e., $\pi_1(X) = \{A_1\}$, then the conclusion follows by Theorem 6.17.

So, suppose that $h > 1$, and that the theorem holds for all sets of fat points Z' with $|\pi_1(\text{supp}(Z'))| < h$. For each $A_i \in \pi_1(X)$, we let $I(Z_{1,A_i})$ denote the ideal that defines the set of fat points $Z_{1,A_i} := \{(P_{i,j_1}; m_{i,j_1}), (P_{i,j_2}; m_{i,j_2}), \cdots, (P_{i,j_{h_i}}; m_{i,j_{h_i}})\}$, i.e., the set of fat points which consists of all the fat points of Z on the horizontal ruling H_{A_i}. We set

$$I(Y_1) = \bigcap_{i=1}^{h-1} I(Z_{1,A_i}) \quad \text{and} \quad I(Y_2) = I(Z_{1,A_h}).$$

The ideals $I(Y_1)$ and $I(Y_2)$ are the defining ideals of a set of fat points in $\mathbb{P}^1 \times \mathbb{P}^1$ with $|\pi_1(\text{supp}(Y_i))| < h$ for $i = 1, 2$. We shall also require the following result about $I(Y_1)$ and $I(Y_2)$.

Claim. For any $j \in \mathbb{N}$, if $i \geq l = l_1 + \cdots + l_h - 1$, then $(I(Y_1) + I(Y_2))_{i,j} = R_{i,j}$.

Proof of the Claim. It is enough to show that $(I(Y_1) + I(Y_2))_{l-1,0} = R_{l-1,0}$. Recall that for each $A_i \in \pi_1(X)$, the integer l_i is defined to be $l_i = \max\{m_{i,j_1}, \ldots, m_{i,j_{v_i}}\}$ where Z_{1,A_i} is as above. Then $I_{Z_{1,A_i}} = \bigcap_{c=1}^{v_i} (H_{A_i}, V_{B_{jc}})^{m_{i,jc}}$. Note that $\deg H_{A_i} = (1,0)$ and $\deg V_{B_{jc}} = (0,1)$. From this description of $I_{Z_{1,A_i}}$, it follows that $H_{A_i}^{l_i} \in I_{Z_{1,A_i}}$. Thus $H_{A_1}^{l_1} \cdots H_{A_{h-1}}^{l_{h-1}} \in I(Y_1)$ and $H_{A_h}^{l_h} \in I(Y_2)$.

Set $J = (H_{A_1}^{l_1} \cdots H_{A_{h-1}}^{l_{h-1}}, H_{A_h}^{l_h}) \subseteq I(Y_1) + I(Y_2)$. Because J is generated by a regular sequence, the bigraded resolution of J is given by the Koszul resolution (see Lemma 2.26):

$$0 \longrightarrow R(-l,0) \longrightarrow R(-l+l_h,0) \oplus R(-l_h,0) \longrightarrow J \longrightarrow 0.$$

If we use this exact sequence to calculate the dimension of $J_{l-1,0}$, then we find

$$\begin{aligned}\dim_k J_{l-1,0} &= (l-1-(l-l_h)+1)+(l-1-l_h+1)-(l-1-l+1)\\ &= l_h+l-l_h = l = \dim_k R_{l-1,0}.\end{aligned}$$

Because we have $J_{l-1,0} \subseteq (I(Y_1)+I(Y_2))_{l-1,0} \subseteq R_{l-1,0}$, it follows that $(I(Y_1)+I(Y_2))_{l-1,0} = R_{l-1,0}$. □

From the short exact sequence

$$0 \longrightarrow I(Y_1)\cap I(Y_2) = I(Z) \longrightarrow I(Y_1)\oplus I(Y_2) \longrightarrow I(Y_1)+I(Y_2) \longrightarrow 0$$

we deduce that

$$\dim_k I(Z)_{i,j} = \dim_k I(Y_1)_{i,j} + \dim_k I(Y_2)_{i,j} - \dim_k (I(Y_1)+I(Y_2))_{i,j}$$

for all $(i,j) \in \mathbb{N}^2$. Thus, when $i \geq l-1$, the above claim implies that

$$\begin{aligned}H_Z(i,j) &= (i+1)(j+1) - \dim_k I(Y_1)_{i,j} - \dim_k I(Y_2)_{i,j} + \dim_k (I(Y_1)+I(Y_2))_{i,j}\\ &= (i+1)(j+1) - \dim_k I(Y_1)_{i,j} + (i+1)(j+1) - \dim_k I(Y_2)_{i,j}\\ &= H_{Y_1}(i,j) + H_{Y_2}(i,j).\end{aligned}$$

Recall that in the construction of α_Z, we have

$$\tilde{\alpha}_Z = (\alpha_{A_1},\dots,\alpha_{A_h}) = (a_{1,0},\dots,a_{1,l_1-1},a_{2,0},\dots,a_{2,l_2-1},\dots,a_{h,0},\dots,a_{h,l_h-1}).$$

Furthermore, from our construction of Y_1 and Y_2 we have $\tilde{\alpha}_{Y_1} = (\alpha_{A_1},\dots,\alpha_{A_{h-1}})$ and $\tilde{\alpha}_{Y_2} = (\alpha_{A_h})$. Since α_Z is simply $\tilde{\alpha}_Z$ reordered in a non-increasing order (and the same for α_{Y_1} and α_{Y_2}), we have

$$\begin{aligned}(\alpha_Z)^*_j &= |\{\alpha_i \in \alpha_Z \mid \alpha_i \geq j\}|\\ &= |\{\alpha_i \in \tilde{\alpha}_Z \mid \alpha_i \geq j\}|\\ &= |\{\alpha_i \in \tilde{\alpha}_{Y_1} \mid \alpha_i \geq j\}| + |\{\alpha_i \in \tilde{\alpha}_{Y_2} \mid \alpha_i \geq j\}|\\ &= |\{\alpha_i \in \alpha_{Y_1} \mid \alpha_i \geq j\}| + |\{\alpha_i \in \alpha_{Y_2} \mid \alpha_i \geq j\}| = (\alpha_{Y_1})^*_j + (\alpha_{Y_2})^*_j.\end{aligned}$$

To complete the proof, for any $j \in \mathbb{N}$ and any $i \geq l-1$, we then have

$$\begin{aligned}H_Z(i,j) &= H_{Y_1}(i,j) + H_{Y_2}(i,j)\\ &= ((\alpha_{Y_1})^*_1 + \cdots + (\alpha_{Y_1})^*_{j+1}) + ((\alpha_{Y_1})^*_1 + \cdots + (\alpha_{Y_1})^*_{j+1}) \quad \text{(induction)}\\ &= (\alpha_Z)^*_1 + \cdots + (\alpha_Z)^*_{j+1} \quad \text{(above identity)}\end{aligned}$$

thus completing the proof.

6.4 ACM fat points in $\mathbb{P}^1 \times \mathbb{P}^1$

As in Chapter 4, we can classify when the quotient ring $R/I(Z)$ is Cohen-Macaulay when Z is a collection of fat points. In fact, Theorem 6.21 can be seen as a direct generalization of Theorem 4.11. As in previous chapters, we work out all the relevant details. We begin with the following definition.

Definition 6.19. A set of fat points $Z \subseteq \mathbb{P}^1 \times \mathbb{P}^1$ is *arithmetically Cohen-Macaulay* (ACM) if $R/I(Z)$ is a Cohen-Macaulay ring.

As in the case of reduced points (see Theorem 4.6), we can find a regular sequence of length two in $R/I(Z)$ when Z is ACM with some additional information about the bidegrees of this regular sequence.

Theorem 6.20. *Let $Z \subseteq \mathbb{P}^1 \times \mathbb{P}^1$ be a set of fat points. If Z is ACM, then there exist elements $\overline{L}_1, \overline{L}_2$ in $R/I(Z)$ such that $L_1 \in R_{1,0}$ and $L_2 \in R_{0,1}$, and L_1, L_2 give rise to a regular sequence in $R/I(Z)$.*

Proof. Because K-dim$R/I(Z) = 2$ (Lemma 6.9) and Z is ACM it follows that there exists a regular sequence of length two in $R/I(Z)$. It is therefore sufficient to show that the elements in the regular sequence have the appropriate degrees. By Lemma 6.7, there exists $L_1 \in R_{1,0}$ such that $\overline{L}_1$ is a nonzero-divisor on $R/I(Z)$. So, it is enough to show that we can find an element of $R_{0,1}$ that is a nonzero-divisor on $R/(I(Z),L_1)$. Note that $R/(I(Z),L_1)$ is a Cohen-Macaulay ring with Krull dimension one (see, for example [98, Lemma 1.3.10] which shows that both the depth and dimension of an R-module M drop by one when we pass to M/zM when z is a regular element).

Let $(I(Z),L_1) = Q_1 \cap \cdots \cap Q_s$ be the primary decomposition of $(I(Z),L_1)$ and set $\wp_l := \sqrt{Q_l}$ for each $l = 1,\ldots,s$. We claim that $(x_0,x_1) \subseteq \wp_l$ for each l. Indeed, because $\overline{L}_1$ is a nonzero-divisor, we have the following bigraded short exact sequence:

$$0 \longrightarrow (R/I(Z))(-1,0) \xrightarrow{\times \overline{L}_1} R/I(Z) \longrightarrow R/(I(Z),L_1) \longrightarrow 0.$$

Thus, $H_{R/(I(Z),L_1)}(i,j) = H_Z(i,j) - H_Z(i-1,j)$ for all $(i,j) \in \mathbb{N}^2$. By Theorem 6.18, if $i \gg 0$, $H_Z(i,0) = H_Z(i-1,0)$, and hence, $H_{R/(I(Z),L_1)}(i,0) = 0$. This implies $(I(Z),L_1)_{i,0} = R_{i,0} = \left[(x_0,x_1)^i\right]_{i,0}$. So, $(x_0,x_1)^i \subseteq Q_l$ for $i \gg 0$ and for each $l = 1,\ldots,s$. Therefore, $(x_0,x_1) \subseteq \wp_l$ for each l.

Now the set of zero-divisors of $R/(I(X),L_1)$ is given by

$$Z(R/(I(X),L_1)) = \bigcup_{i=1}^{s} \wp_i.$$

Suppose that there is no nonzero-divisor of degree $(0,1)$, and consequently, $R_{0,1} \subseteq Z(R/(I(Z),L_1)$. Thus $R_{0,1} = \bigcup_{i=1}^{s} (\wp_i)_{0,1}$. But $R_{0,1}$ is an infinite vector space that cannot be written as a union of smaller vector spaces unless there is some i such that $(\wp_i)_{0,1} = R_{0,1}$. We have just shown that $(x_0,x_1) \subseteq \wp_j$ for all j. Thus $\mathbf{m} = (x_0,x_1,y_0,y_1) \subseteq \wp_i$, that is, the maximal ideal $\mathbf{m}$ of R, is an associated prime of $(I(Z),L_1)$. But then every homogeneous element of $R/(I(Z),L_1)$ is a zero-divisor, contradicting the fact that $R/(I(Z),L_1)$ is a Cohen-Macaulay ring with Krull dimension one. So $R/(I(Z),L_1)$ has a nonzero-divisor of degree $(0,1)$.

We now present the following characterization of ACM fat points in $\mathbb{P}^1 \times \mathbb{P}^1$.

Theorem 6.21. *Let $Z \subseteq \mathbb{P}^1 \times \mathbb{P}^1$ be a set of fat points with Hilbert function H_Z. Then the following are equivalent:*

(i) Z is arithmetically Cohen-Macaulay.
(ii) ΔH_Z is the Hilbert function of a bigraded artinian quotient of $k[x_1,y_1]$.
(iii) $\alpha_Z^* = \beta_Z$.
(iv) The set S_Z has no incomparable elements with respect to the partial order $\succeq$ on $\mathbb{N}^\nu$.

Proof. The equivalence of (iii) and (iv) is Theorem 6.16. We prove the remaining equivalences.

$(i) \Rightarrow (ii)$ Let L_1,L_2 be the regular sequence of Theorem 6.20. By making a linear change of coordinates in the x_i's, and a linear change of coordinates in the y_i's, we can assume that $L_1 = x_0$ and $L_2 = y_0$ give rise to a regular sequence in $R/I(Z)$.

From the short exact sequences

$$\begin{array}{ccccccc} 0 \to & (R/I(Z))(-1,0) & \stackrel{\times \overline{x}_0}{\to} & R/I(Z) & \to & R/(I(Z),x_0) & \to 0 \\ 0 \to & (R/(I(Z),x_0))(0,-1) & \stackrel{\times \overline{y}_0}{\longrightarrow} & R/(I(Z),x_0) & \to & R/(I(Z),x_0,y_0) & \to 0 \end{array}$$

it follows that $H_{R/(I(Z),x_0,y_0)}(i,j) = \Delta H_Z(i,j)$ for all $(i,j) \in \mathbb{N}^2$. Moreover,

$$R/(I(Z),x_0,y_0) \cong \frac{R/(x_0,y_0)}{(I(Z),x_0,y_0)/(x_0,y_0)} \cong k[x_1,y_1]/J$$

where J is a bihomogeneous ideal with $J \cong (I(Z),x_0,y_0)/(x_0,y_0)$. By Theorem 6.18 it follows that $\Delta H_Z(i,j) = 0$ if $i \gg 0$ or $j \gg 0$. Hence $k[x_1,y_1]/J$ is a bigraded artinian ring.

$(ii) \Rightarrow (iii)$ (This argument is almost word-for-word the same as the argument used in Theorem 4.11.) Suppose that ΔH_Z is the Hilbert function of a bigraded artinian quotient of $k[x_1,y_1]$. Because $\dim_k k[x_1,y_1]_{i,j} = 1$ for all (i,j), $\Delta H_Z(i,j) = 1$ or 0. If we write ΔH_Z as an infinite matrix whose indexing starts from zero, rather than one, then we have

$\Delta H_Z =$

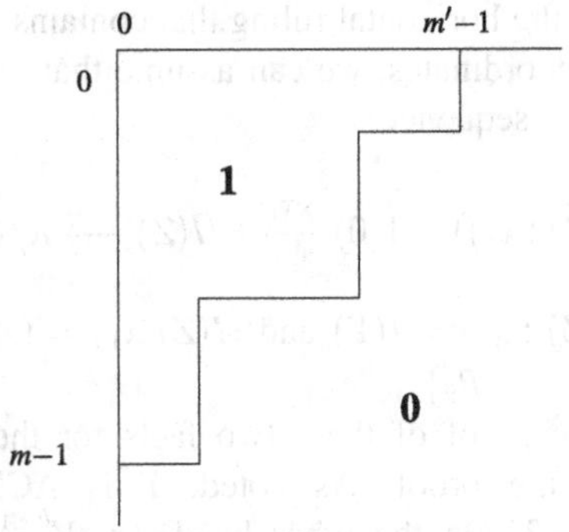

We can recover the Hilbert function H_Z from ΔH_Z using the fact that

$$H_Z(i,j) = \sum_{(0,0)\preceq(k,l)\preceq(i,j)} \Delta H_Z(k,l).$$

Comparing this formula to Theorem 6.18, we have

$$\begin{aligned}
H_Z(m-1,0) &= m = \alpha_1^* &&= \Delta H_Z(0,0)+\Delta H_Z(1,0)+\cdots+\Delta H_Z(m-1,0)\\
H_Z(m-1,1) &= \alpha_1^*+\alpha_2^* &&= \alpha_1^*+\Delta H_Z(0,1)+\Delta H_Z(1,1)+\cdots+\Delta H_X(m-1,1)\\
H_X(m-1,2) &= \alpha_1^*+\alpha_2^*+\alpha_3^* &&= \alpha_1^*+\alpha_2^*+\Delta H_Z(0,2)+\Delta H_Z(1,2)+\cdots+\Delta H_Z(m-1,2)\\
&\vdots
\end{aligned}$$

In other words,

$$\alpha_j^* = \Delta H_Z(0,j-1)+\cdots+\Delta H_Z(m-1,j-1)$$

counts the number of 1's that appears in the $(j-1)$-th column of ΔH_Z. By a similar argument, β_j^* counts the number of 1's that appear in the $(j-1)$-th row of ΔH_Z.

Because ΔH_Z consists of only 1's and 0's, we can view ΔH_Z as a Ferrers diagram. But then it follows that $(\alpha_1^*,\ldots,\alpha_{m'}^*)^* = \beta_Z^*$, that is, $\alpha_Z = \beta_Z^*$.

$(iv) \Rightarrow (i)$ We will give a proof by induction on $|S_Z|$. If $|S_Z| = 1$, then $S_Z = \{(\underbrace{1,\ldots,1}_{v})\}$, and so Z consists of v reduced points on a horizontal ruling. But then Z is ACM by Lemma 4.8.

Now suppose that $|S_Z| > 1$. Because S_Z has no incomparable elements with respect to the partial order $\succeq$ on $\mathbb{N}^v$, it has a unique maximal element, say $(m_{k,1},\ldots,m_{k,v})$. As in the proof of Theorem 6.16, we construct the set of fat points given by $Y = \{(P_{i,j}; m'_{i,j}) \mid 1 \le i \le h,\ 1 \le j \le v\}$ where

$$m'_{i,j} = \begin{cases} m_{i,j} & \text{if } i \neq k \\ m_{i,j}-1 & \text{if } i = k. \end{cases}$$

By our construction Y has $S_Y = S_Z \setminus \{(m_{k,l},\ldots,m_{k,v})\}$. Furthermore, S_Y has no incomparable elements with respect to $\succeq$. Because $|S_Y| < |S_Z|$, by induction we know that Y is ACM.

Let $H = H_{A_k}$ be the horizontal ruling that contains the points $P_{k,1}, P_{k,2}, \ldots, P_{k,v}$. After a change of coordinates, we can assume that $H = x_1$. We can construct the following short exact sequence

$$0 \longrightarrow (R/(I(Z) : x_1))(-1,0) \xrightarrow{\times \overline{x}_1} R/I(Z) \longrightarrow R/(I(Z), x_1) \longrightarrow 0. \tag{6.1}$$

We claim that $(I(Z) : x_1) = I(Y)$ and $(I(Z), x_1) = (x_1, V_{B_1}^{m_{k,1}} V_{B_2}^{m_{k,2}} \cdots V_{B_v}^{m_{k,v}})$ where $\pi_2(\mathrm{Supp}(Z)) = \{B_1, \ldots, B_v\}$.

We postpone the proof of these two facts for the moment, and first explain how to complete the proof. As noted, Y is ACM by induction, and thus proj-dim$(R/I(Y)) = 2$. On the other hand, $(x_1, V_{B_1}^{m_{k,1}} V_{B_2}^{m_{k,2}} \cdots V_{B_v}^{m_{k,v}})$ is generated by a regular sequence. So, the ideal $(I(Z), x_1)$ is a complete intersection, and thus proj-dim$(R/(I(Z), x_1) = 2$. Lemma 2.23 thus implies that proj-dim$(R/I(Z)) \leq 2$. By the Auslander-Buchsbaum Formula (see Theorem 2.22)

$$2 \geq \text{proj-dim}(R/I(Z)) = 4 - \mathrm{depth}(R/I(Z)) \geq 2$$

since $1 \leq \mathrm{depth}(R/I(Z)) \leq 2$. So proj-dim$(R/I(Z)) = 2$. Consequently, we have that depth$(R/I(Z)) = 2$, which implies that Z is ACM, as desired.

So it suffices to proof the following two claims.

Claim. $(I(Z) : x_1) = I(Y)$

Proof of the Claim. Note that

$$(I(Z) : x_1) = \left(\bigcap_{i=1}^{h} \left(\bigcap_{j=1}^{v} I(P_{i,j})^{m_{i,j}} \right) \right) : (x_1) = \bigcap_{i=1}^{h} \left(\bigcap_{j=1}^{v} (I(P_{i,j})^{m_{i,j}} : x_1) \right).$$

Now, if $i \neq k$, then $(I(P_{i,j})^{m_{i,j}} : x_1) = I(P_{i,j})^{m_{i,j}}$ because x_1 does not pass through the point $P_{i,j}$. On the other hand, if $i = k$, then $(I(P_{i,j})^{m_{i,j}} : x_1) = I(P_{i,j})^{m_{i,j}-1}$ by Lemma 6.6. So,

$$(I(Z) : x_1) = \bigcap_{\substack{i=1 \\ i \neq k}}^{h} \left(\bigcap_{j=1}^{v} I(P_{i,j})^{m_{i,j}} \right) \cap \left(\bigcap_{j=1}^{v} I(P_{k,j})^{m_{k,j}-1} \right) = I(Y).$$

This completes the proof of the claim.

Claim. $(I(Z), x_1) = (x_1, V_{B_1}^{m_{k_1}} V_{B_2}^{m_{k,2}} \cdots V_{B_v}^{m_{k,v}})$.

Proof of the Claim. Let $I(P_{i,j})^{m_{i,j}} = (H_{A_i}, V_{B_j})^{m_{i,j}}$ be the defining ideal of a fat point of Z. Because $(m_{k,1}, \ldots, m_{k,v})$ is the maximal element of S_Z, this means that $m_{i,j} \leq m_{k,j}$ for any i. But then $V_{B_j}^{m_{k,j}} \in I(P_{i,j})^{m_{i,j}}$ and thus $V_{B_1}^{m_{k,1}} V_{B_2}^{m_{k,2}} \cdots V_{B_v}^{m_{k,v}} \in I(Z)$. This shows that $(x_1, V_{B_1}^{m_{k,1}} V_{B_2}^{m_{k,2}} \cdots V_{B_v}^{m_{k,v}}) \subseteq (I(Z), x_1)$.

For the reverse inclusion, let $F \in (I(Z), x_1)$. Assume, without loss of generality, that $\deg F = (a,b)$. If we divide F by x_1, by the Division Algorithm we have $F = F'x_1 + F''$ where no term of F'' is divisible by x_1. Because F is bihomogeneous, this implies that $F'' = x_0^a G$ where G is a polynomial of degree $(0,b)$ only in the variables y_0 and y_1. Furthermore, since $F \in (I(Z), x_1)$, we have that $x_0^a G \in I(Z)$. In particular, for each $j = 1, \ldots, v$, we have

$$x_0^a G \in I(P_{k,j})^{m_{k,j}} = (x_1, V_{B_k})^{m_{k,j}} = (x_1^{m_{k,j}}, x_1^{m_{k,j}-1} V_{B_k}, \ldots, V_{B_j}^{m_{k,j}}).$$

So, there exist $H_0, \ldots, H_{m_{k,j}} \in R$ such that

$$x_0^a G = H_0 x_1^{m_{k,j}} + H_1 x_1^{m_{k,j}-1} V_{B_k} + \cdots + H_{m_{k,j}-1} x_1 V_{B_j}^{m_{k,j}-1} + H_{m_{k,j}} V_{B_j}^{m_{k,j}}$$

If we divide $H_{m_{k,j}}$ by x_1 using the Division Algorithm, we have $H_{m_{k,j}} = G_1 x_1 + G_2$ where x_1 does not divide any term of G_2. Because x_1 does not divide any term of $x_0^a G$, we must have

$$x_0^a G = \underbrace{H_0 x_1^{m_{k,j}} + H_1 x_1^{m_{k,j}-1} V_{B_k} + \cdots + H_{m_{k,j}-1} x_1 V_{B_j}^{m_{k,j}-1} + G_1 x_1 V_{B_j}^{m_{k,j}}}_{=0} + G_2 V_{B_j}^{m_{k,j}} = G_2 V_{B_j}^{m_{k,j}},$$

or in other words, $V_{B_j}^{m_{k,j}}$ divides G. Because this is true for all j, G must be divisible by $V_{B_1}^{m_{k,1}} \cdots V_{B_v}^{m_{k,v}}$, so the conclusion of the second claim now follows.

Example 6.22. The running example of Section 6.2 (see Example 6.11) is not ACM because there exist incomparable elements in S_Z. Indeed, as we showed in Example 6.14, the elements $(4,2,0)$ and $(0,0,1)$ both belong to S_Z. But these two elements are incomparable with respect to the partial order $\succeq$.

Example 6.23. Consider the set of fat points in $\mathbb{P}^1 \times \mathbb{P}^1$ given in Figure 6.2.
For this set of fat points, we have

$$\alpha_Z = (12,10,6,6,5,3,2,2,1,1) \text{ and } \beta_Z = (10,6,6,5,4,4,3,3,2,2,2,1).$$

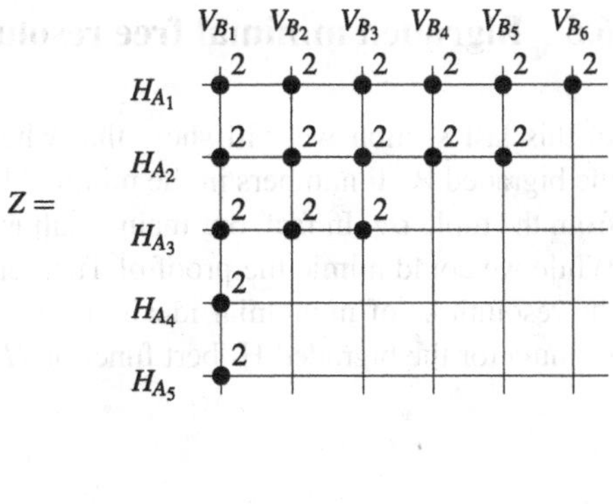

Fig. 6.2 A non-ACM set of fat points in $\mathbb{P}^1 \times \mathbb{P}^1$

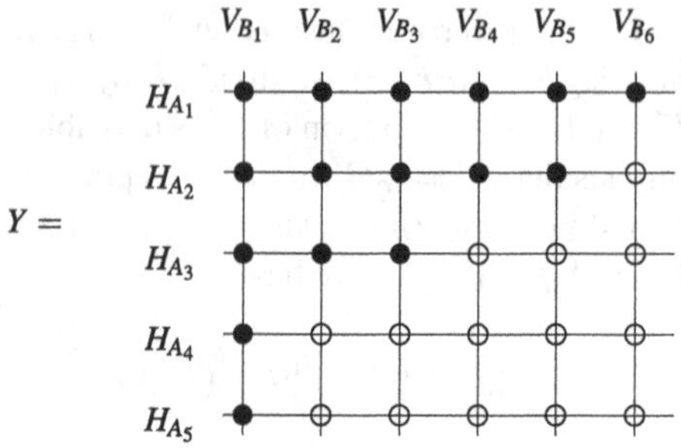

Fig. 6.3 An ACM set of fat points in $\mathbb{P}^1 \times \mathbb{P}^1$

It then follows that Z is not ACM because $\alpha_Z^* = (10,8,6,5,5,4,2,2,2,2,1,1) \neq \beta_Z$. We can also see this from S_Z. For this example,

$$S_Z = \{(2,2,2,2,2,2),(1,1,1,1,1,1),(2,2,2,2,2,0),(1,1,1,1,1,0),(2,2,2,0,0,0)\\ (1,1,1,0,0,0),(2,0,0,0,0,0),(1,0,0,0,0,0),(2,0,0,0,0,0),(1,0,0,0,0,0)\}.$$

In particular, $(1,1,1,1,1,1)$ and $(2,0,0,0,0,0)$ are incomparable. Note that in this example the support of Z is an ACM set of points.

Example 6.24. We consider another set of fat points Y in $\mathbb{P}^1 \times \mathbb{P}^1$, as given in Figure 6.3. In the figure, a solid dot represents a point of multiplicity two, and an empty dot represents a point of multiplicity one.

The set Y is an ACM set of fat points in $\mathbb{P}^1 \times \mathbb{P}^1$ by Theorem 6.21 because

$$\alpha_Y = (12,11,9,7,7,6,5,3,1,1) \text{ and } \alpha_Y^* = \beta_Y = (10,8,8,7,7,6,5,3,3,2,2,1).$$

In this case, the set S_Z is totally ordered by $\succeq$:

$$S_Z = \{(2,2,2,2,2,2),(2,2,2,2,2,1),(2,2,2,1,1,1),(2,1,1,1,1,1),(2,1,1,1,1,1),\\ (1,1,1,1,1,1),(1,1,1,1,1,0),(1,1,1,0,0,0),(1,0,0,0,0,0),(1,0,0,0,0,0)\}.$$

6.5 Bigraded minimal free resolution of ACM fat points

In this last section we will show that when Z is an ACM set of fat points, then the bigraded Betti numbers in the minimal free resolution can be computed directly from the tuple α_Z. In fact, our main result is a direct generalization of Theorem 5.3. While we could mimic the proof of Theorem 5.3, we present a new proof that uses the resolutions of monomial ideals in two variables. As a corollary, we present a formula for the bigraded Hilbert function H_Z in terms of α_Z or S_Z.

We begin with a lemma that enables us to reduce the computation of the bigraded Betti numbers of $I(Z)$ to the computation of the bigraded Betti numbers of $(I(Z)+L)/(L)$ in $R/(L)$ where L is a regular element.

Lemma 6.25. *Let I be a bihomogeneous ideal of R and suppose that L is a bihomogeneous nonzero-divisor on R/I. Set $S=R/(L)$. If $J=(I+L)/(L)\subseteq S$, then*

$$\beta^R_{i,(j_1,j_2)}(R/I)=\beta^S_{i,(j_1,j_2)}(S/J) \text{ for all } i\in\mathbb{N} \text{ and } (j_1,j_2)\in\mathbb{N}^2.$$

Proof. By applying [79, Corollary 20.4], we have

$$\operatorname{Tor}^R_i(R/I,k)=\operatorname{Tor}^S_i(S/J,k) \text{ for all } i\in\mathbb{N}.$$

Because I and J are bigraded ideals, then the above Tor-modules are also bigraded. In particular,

$$\beta^R_{i,(j_1,j_2)}(R/I)=\dim_k \operatorname{Tor}^R_i(R/I,k)_{(j_1,j_2)}=\dim_k \operatorname{Tor}^S_i(S/J,k)_{(j_1,j_2)}=\beta^S_{i,(j_1,j_2)}(S/J).$$

We also require the following result about the bigraded minimal free resolution of monomial ideals in $k[x,y]$.

Theorem 6.26. *Let $I=\langle x^{a_1}y^{b_1},x^{a_2}y^{b_2},\ldots,x^{a_s}y^{b_s}\rangle$ be a monomial ideal in $R=k[x,y]$ such that $a_1>a_2>\cdots>a_s\geq 0$ and $0\leq b_1<b_2<\cdots<b_s$. If R is bigraded, i.e., $\deg x=(1,0)$ and $\deg y=(0,1)$, then the bigraded minimal free resolution of I has the form*

$$0\longrightarrow\bigoplus_{i=1}^{s-1}R(-a_i,-b_{i+1})\longrightarrow\bigoplus_{i=1}^{s}R(-a_i,-b_i)\longrightarrow I\longrightarrow 0.$$

Proof. The resolution is a consequence of [74, Proposition 3.1], but also taking into account the bigraded structure.

We now compute the graded Betti numbers of $I(Z)$ when $Z\subseteq\mathbb{P}^1\times\mathbb{P}^1$ is an ACM set of fat points. Let $\alpha_Z=(\alpha_1,\ldots,\alpha_m)$ be the tuple associated with Z. Define the following two sets from α_Z:

$$C_Z:=\{(m,0),(0,\alpha_1)\}\cup\{(i-1,\alpha_i)\mid \alpha_i-\alpha_{i-1}<0\}$$
$$V_Z:=\{(m,\alpha_m)\}\cup\{(i-1,\alpha_{i-1})\mid \alpha_i-\alpha_{i-1}<0\}.$$

We take $\alpha_{-1}=0$. With this notation, we have:

Theorem 6.27. *Let $Z\subseteq\mathbb{P}^1\times\mathbb{P}^1$ be an ACM set of fat points with $\alpha_Z=(\alpha_1,\ldots,\alpha_m)$. Let C_Z and V_Z be constructed from α_Z as above. Then the bigraded minimal free resolution of $I(Z)$ is given by*

$$0\longrightarrow\bigoplus_{(v_1,v_2)\in V_Z}R(-v_1,-v_2)\longrightarrow\bigoplus_{(c_1,c_2)\in C_Z}R(-c_1,-c_2)\longrightarrow I(Z)\longrightarrow 0.$$

$H_{S/J} = \Delta H_Z =$

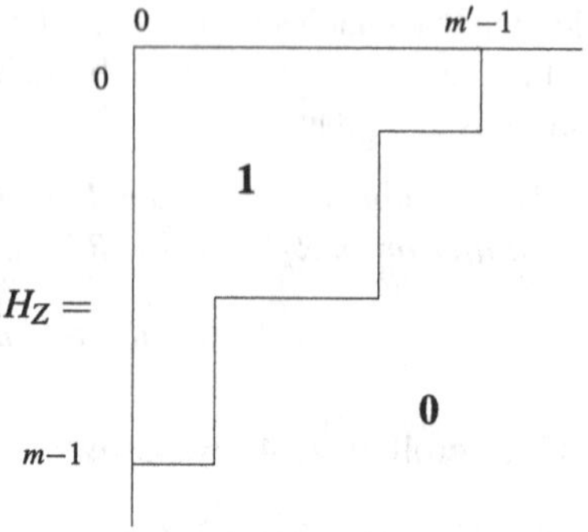

Proof. Because Z is ACM, by Lemma 6.20 there exists a regular sequence L_1, L_2 on $R/I(Z)$ such that $\deg L_1 = (1,0)$ and $\deg L_2 = (0,1)$. After a change of coordinates, we can assume that $L_1 = x_0$ and $L_2 = y_0$.

By applying Lemma 6.25 twice, we have

$$\beta^R_{i,(j_1,j_2)}(R/I(Z)) = \beta^S_{i,(j_1,j_2)}(S/J)$$

where $S = k[x_1,y_1] \cong R/(x_0,y_0)$ and $J \cong (I(Z),x_0,y_0)/(x_0,y_0)$. Because J is a bihomogeneous ideal of S, J is a monomial ideal of S (in fact, J is an artinian monomial ideal). So, it suffices to compute the bigraded Betti numbers of S/J. By Theorem 6.26, it thus suffices to find the degrees of the minimal generators of J.

As in the proof of Theorem 6.21, we know that the Hilbert function of S/J has the form. As noted in the proof of Theorem 6.21, the number of ones in the $(i-1)$-th row of this matrix is the i-th entry of β^*_Z. However, because Z is ACM, the i-th entry of β^*_Z is the same as the i-th entry of α_Z.

The degrees of the minimal monomial generators of J correspond to the minimal $(a,b) \in \mathbb{N}^2$ such that $\Delta H_Z(a,b) = 0$ with respect to the natural partial order $\succeq$ on $\mathbb{N}^2$. In particular, the set of the degrees of the minimal monomial generators of J is precisely the set C_Z. We can then compute the bigraded minimal free resolution of $I(Z)$ using Theorem 6.26. The degrees of the minimal syzygies will then be the elements of the set V_Z.

Corollary 6.28. *Let $Z \subseteq \mathbb{P}^1 \times \mathbb{P}^1$ be an ACM set of fat points with $\alpha_Z = (\alpha_1, \ldots, \alpha_m)$, then the Hilbert function of Z is*

$$H_Z = \begin{bmatrix} 1 & 2 & \cdots & \alpha_1 - 1 & \alpha_1 & \alpha_1 & \cdots \\ 1 & 2 & \cdots & \alpha_1 - 1 & \alpha_1 & \alpha_1 & \cdots \\ \vdots & \vdots & & \vdots & \vdots & \vdots & \ddots \end{bmatrix} + \begin{bmatrix} 0 & 0 & \cdots & 0 & 0 & 0 & \cdots \\ 1 & 2 & \cdots & \alpha_2 - 1 & \alpha_2 & \alpha_2 & \cdots \\ 1 & 2 & \cdots & \alpha_2 - 1 & \alpha_2 & \alpha_2 & \cdots \\ \vdots & \vdots & & \vdots & \vdots & \vdots & \ddots \end{bmatrix}$$

$$+\cdots+\begin{bmatrix} 0 & 0 & \cdots & 0 & 0 & 0 & \cdots \\ \vdots & \vdots & & \vdots & \vdots & \vdots & \\ 0 & 0 & \cdots & 0 & 0 & 0 & \cdots \\ 1 & 2 & \cdots & \alpha_m - 1 & \alpha_m & \alpha_m & \cdots \\ 1 & 2 & \cdots & \alpha_m - 1 & \alpha_m & \alpha_m & \cdots \\ \vdots & \vdots & & \vdots & \vdots & \vdots & \ddots \end{bmatrix}.$$

Proof. We sketch out two different ways to prove this result, and leave the fine details to the reader.

The first way to prove this statement is to note that the bigraded minimal free resolution given in Theorem 6.27 implies that for all $(i,j) \in \mathbb{N}^2$,

$$\dim_k(I(Z))_{i,j} = \sum_{(c_1,c_2)\in C_Z} (\dim_k R(-c_1,-c_2)_{i,j}) - \sum_{(v_1,v_2)\in V_Z} (\dim_k R(-v_1,-v_2)_{i,j})$$

We can now compute $H_Z(i,j) = \dim_k R_{i,j} - \dim_k(I(Z))_{i,j}$ using the above identity and the fact that

$$\dim_k R(-a,-b)_{i,j} = \begin{cases} (i-a+1)(j-b+1) & \text{if } (i,j) \succeq (a,b) \\ 0 & \text{otherwise.} \end{cases}$$

Alternatively, one can use the short exact sequence (6.1) from Theorem 6.21, that is,

$$0 \longrightarrow (R/(I(Z):x_1))(-1,0) \xrightarrow{\times \bar{x}_1} R/I(Z) \longrightarrow R/(I(Z),x_1) \longrightarrow 0,$$

to prove the statement using induction. One will also need to use the two claims found in Theorem 6.21.

Like the case of simple points, there is an alternative way to express the Hilbert function of a set of fat points in $\mathbb{P}^1 \times \mathbb{P}^1$ using the tuple S_Z. Although we do not present the proof, the key is that when Z is ACM, the sum of the coordinates of the i-th entry of S_Z is equal to α_i of α_Z.

Corollary 6.29. *Let $Z \subseteq \mathbb{P}^1 \times \mathbb{P}^1$ be an ACM set of fat points with $S_Z = \{s_1,\ldots,s_r\}$ where $s_i = (s_{i,1},\ldots,s_{i,v}) \in \mathbb{N}^v$ for $i = 1,\ldots,r$. Because S_Z has no incomparable elements, we can assume that $s_1 \succeq s_2 \succeq \cdots \succeq s_r$. Let $t_i = |s_i| = \sum_{k=1}^{v} s_{i,k}$ for $i = 1,\ldots,r$. Then*

$$\Delta H_Z = \begin{bmatrix} \underbrace{1\ 1\ \cdots\ 1}_{t_1}\ 1\ 0\ 0\ \cdots \\ \underbrace{1\ \cdots\ 1}_{t_2}\ 1\ 0\ 0\ 0\ \cdots \\ \vdots\ \vdots\ \ \vdots\ \vdots\ \vdots\ \vdots\ \vdots\ \cdots \\ \underbrace{1\ \cdots\ 1}_{t_r}\ 0\ 0\ 0\ 0\ \cdots \\ 0\ 0\ \ 0\ 0\ 0\ 0\ 0\ \cdots \\ \vdots\ \vdots\ \ \vdots\ \vdots\ \vdots\ \vdots\ \vdots\ \ddots \end{bmatrix}.$$

Example 6.30. When $Z \subseteq \mathbb{P}^1 \times \mathbb{P}^1$ is an ACM set of fat points, then ΔH_Z is the Hilbert function of a bigraded artinian quotient of $k[x_1, y_1]$, and furthermore, by the Corollary 6.29, we can even write down ΔH_Z directly from S_Z. As we showed in Chapter 5, if we start from any Hilbert function H of any bigraded artinian quotient of $k[x_1, y_1]$, we can find an ACM set of reduced points X such that $\Delta H_X = H$. However, this is no longer the case when Z is a set of fat points. As noted in [43, Remark 2.4], the following Hilbert function H is the Hilbert function of a bigraded artinian quotient of $k[x_1, y_1]$,

$$H = \begin{bmatrix} 1\ 1\ 1\ 1\ 0\ \cdots \\ 1\ 1\ 1\ 0\ 0\ \cdots \\ 1\ 1\ 0\ 0\ 0\ \cdots \\ 0\ 0\ 0\ 0\ 0\ \cdots \\ \vdots\ \vdots\ \vdots\ \vdots\ \vdots\ \ddots \end{bmatrix}$$

but we can easily check that there is no set Z of fat points in $\mathbb{P}^1 \times \mathbb{P}^1$ with some $m_{ij} > 1$ which has $\Delta H_Z = H$ as its first difference Hilbert function.

Example 6.31. Let Y be the ACM set of fat points of Example 6.24. As computed in this example, $\alpha_Y = (12, 11, 9, 7, 7, 6, 5, 3, 1, 1)$. So

$$C_Y := \{(10,0), (8,1), (7,3), (6,5), (5,6), (3,7), (2,9), (1,11), (0,12)\}$$
$$V_Y := \{(10,1), (8,3), (7,5), (6,6), (5,7), (3,9), (2,11), (1,12)\}.$$

The sets C_Y and V_Y then give the shifts in the bigraded minimal free resolution of $I(Y)$ as described by Theorem 6.27. Furthermore, the Hilbert function of H_Y is given by

$$H_Y = \begin{bmatrix} 1 & 2 & 3 & 4 & 5 & 6 & 7 & 8 & 9 & 10 & 11 & 12 & 12 & 12 & \cdots \\ 2 & 4 & 6 & 8 & 10 & 12 & 14 & 16 & 18 & 20 & 22 & 23 & 23 & 23 & \cdots \\ 3 & 6 & 9 & 12 & 15 & 18 & 21 & 24 & 27 & 29 & 31 & 32 & 32 & 32 & \cdots \\ 4 & 8 & 12 & 16 & 20 & 24 & 28 & 31 & 34 & 36 & 38 & 39 & 39 & 39 & \cdots \\ 5 & 10 & 15 & 20 & 25 & 30 & 35 & 38 & 41 & 43 & 45 & 46 & 46 & 46 & \cdots \\ 6 & 12 & 18 & 24 & 30 & 36 & 41 & 44 & 47 & 49 & 51 & 52 & 52 & 52 & \cdots \\ 7 & 14 & 21 & 28 & 35 & 41 & 46 & 49 & 52 & 54 & 56 & 57 & 57 & 57 & \cdots \\ 8 & 16 & 24 & 31 & 38 & 44 & 49 & 52 & 55 & 57 & 59 & 60 & 60 & 60 & \cdots \\ 9 & 17 & 25 & 32 & 39 & 45 & 50 & 53 & 56 & 58 & 60 & 61 & 61 & 61 & \cdots \\ 10 & 18 & 26 & 33 & 40 & 46 & 51 & 54 & 57 & 59 & 61 & 62 & 62 & 62 & \cdots \\ 10 & 18 & 26 & 33 & 40 & 46 & 51 & 54 & 57 & 59 & 61 & 62 & 62 & 62 & \cdots \\ 10 & 18 & 26 & 33 & 40 & 46 & 51 & 54 & 57 & 59 & 61 & 62 & 62 & 62 & \cdots \\ \vdots & \vdots & \vdots & \vdots & \vdots & \vdots & \vdots & \vdots & \vdots & \vdots & \vdots & \vdots & \vdots & \vdots & \ddots \end{bmatrix}.$$

We can compute ΔH_Y from the above matrix, but instead, let us use S_Y. For this set of points we have

$$S_Y = \{(2,2,2,2,2,2),(2,2,2,2,2,1),(2,2,2,1,1,1),(2,1,1,1,1,1),(2,1,1,1,1,1),$$
$$(1,1,1,1,1,1),(1,1,1,1,1,0),(1,1,1,0,0,0),(1,0,0,0,0,0),(1,0,0,0,0,0)\}.$$

So, $(t_1,t_2,\ldots,t_r) = \alpha_Y = (12,11,9,7,7,6,5,3,1,1)$. Thus, by Corollary 6.29

$$\Delta H_Y = \begin{bmatrix} 1 & 1 & 1 & 1 & 1 & 1 & 1 & 1 & 1 & 1 & 1 & 1 & 0 & 0 & \cdots \\ 1 & 1 & 1 & 1 & 1 & 1 & 1 & 1 & 1 & 1 & 1 & 0 & 0 & 0 & \cdots \\ 1 & 1 & 1 & 1 & 1 & 1 & 1 & 1 & 1 & 0 & 0 & 0 & 0 & 0 & \cdots \\ 1 & 1 & 1 & 1 & 1 & 1 & 1 & 0 & 0 & 0 & 0 & 0 & 0 & 0 & \cdots \\ 1 & 1 & 1 & 1 & 1 & 1 & 1 & 0 & 0 & 0 & 0 & 0 & 0 & 0 & \cdots \\ 1 & 1 & 1 & 1 & 1 & 1 & 0 & 0 & 0 & 0 & 0 & 0 & 0 & 0 & \cdots \\ 1 & 1 & 1 & 1 & 1 & 0 & 0 & 0 & 0 & 0 & 0 & 0 & 0 & 0 & \cdots \\ 1 & 1 & 1 & 0 & 0 & 0 & 0 & 0 & 0 & 0 & 0 & 0 & 0 & 0 & \cdots \\ 1 & 0 & 0 & 0 & 0 & 0 & 0 & 0 & 0 & 0 & 0 & 0 & 0 & 0 & \cdots \\ 1 & 0 & 0 & 0 & 0 & 0 & 0 & 0 & 0 & 0 & 0 & 0 & 0 & 0 & \cdots \\ 0 & 0 & 0 & 0 & 0 & 0 & 0 & 0 & 0 & 0 & 0 & 0 & 0 & 0 & \cdots \\ 0 & 0 & 0 & 0 & 0 & 0 & 0 & 0 & 0 & 0 & 0 & 0 & 0 & 0 & \cdots \\ \vdots & \vdots & \vdots & \vdots & \vdots & \vdots & \vdots & \vdots & \vdots & \vdots & \vdots & \vdots & \vdots & \vdots & \ddots \end{bmatrix}.$$

6.6 Additional notes

Fat points in $\mathbb{P}^1 \times \mathbb{P}^1$ were initially studied by the first author in her PhD thesis [42] (which appeared as [43]). In particular, the first author introduced the tuple S_Z to classify ACM sets of fat points Z in $\mathbb{P}^1 \times \mathbb{P}^1$ (Theorem 6.21, $(i) \Leftrightarrow (iv)$). Also a description of the Hilbert function of Z in terms of S_Z (Corollary 6.29) can also be found in this work.

The introduction of the tuples α_Z and β_Z to study fat points in $\mathbb{P}^1 \times \mathbb{P}^1$ first appeared in a joint work of the authors [45]. The equivalence between ACM sets of fat points and the equality $\alpha_Z^* = \beta_Z$ first appeared in this paper. The description of the bigraded minimal free resolution of $I(Z)$ (Theorem 6.27) and its Hilbert function in terms of α_Z (Corollary 6.28) also first appeared in this paper.

While the main results of this chapter have been known for a while, some of the proofs presented in this chapter are new. Specifically, the proof of the implication $(iv) \Rightarrow (i)$ in Theorem 6.21 is new. The proof of the bigraded Betti numbers (Theorem 6.27) that we presented here is also new.

In Chapter 3, we introduced the concept of a separator of a point (see Definition 3.34). As we in saw in Theorem 4.11, one can classify ACM sets of reduced points in terms of the separators. An analog of a separator of a fat point in $\mathbb{P}^1 \times \mathbb{P}^1$ was developed by both authors [49, 51]. While we do not know if one can classify ACM sets of fat points in $\mathbb{P}^1 \times \mathbb{P}^1$ in terms of their separators, the paper [51] describes how to compute a generalization of the degree of a separator when Z is an ACM set of fat points in $\mathbb{P}^1 \times \mathbb{P}^1$.

As we saw in Chapter 5, the interpolation problem has been solved for ACM sets of reduced points in $\mathbb{P}^1 \times \mathbb{P}^1$, that is, given a numerical function $H : \mathbb{N}^2 \to \mathbb{N}$, we can decide whether it is the Hilbert function of an ACM set of reduced points in $\mathbb{P}^1 \times \mathbb{P}^1$. However the corresponding problem remains open for fat points in $\mathbb{P}^1 \times \mathbb{P}^1$, even if we restrict to ACM sets of fat points (for example, see Example 6.30, which was first appeared in [43]). This fact should not be too surprising given the fact that the interpolation problem also remains open for fat points in $\mathbb{P}^2$ (see, for example, [35, 58]). We leave this as a problem:

Problem 6.32. *What functions $H : \mathbb{N}^2 \to \mathbb{N}$ can be the Hilbert function of an ACM set of fat points $Z \subseteq \mathbb{P}^1 \times \mathbb{P}^1$?*

Our understanding of the bigraded minimal free resolution of an arbitrary set of fat points in $\mathbb{P}^1 \times \mathbb{P}^1$ is very far from complete. Some partial progress has been made by studying the Castelnuovo-Mumford regularity of these ideals. Roughly speaking, the Castelnuovo-Mumford regularity provides a bound on the degrees of the syzygies that appear in the minimal free resolution, thus giving us some crude information about the resolution. The second author and Hà [56] bounded the Castelnuovo-Mumford regularity of ideals of fat points in $\mathbb{P}^{n_1} \times \cdots \times \mathbb{P}^{n_r}$ by ignoring the multi-graded structure, and only considering the graded structure. The second author and Sidman [89] investigated the multi-graded version of regularity, as defined by Maclagan and Smith [69], of fat points in $\mathbb{P}^{n_1} \times \cdots \times \mathbb{P}^{n_r}$. For ACM sets of fat points, exact values for the multi-graded regularity were given.

The study of fat points in $\mathbb{P}^1 \times \mathbb{P}^1$ has recently received some additional attention in the study of symbolic powers of ideals (see, for example, [3, 54]). We will return to this topic in Chapter 8.

As a final comment, the first author wrote a survey [44] on fat points in $\mathbb{P}^1 \times \mathbb{P}^1$ a number of years ago. Many of the results of this survey also appeared in this chapter.

Chapter 7
Double points and their resolution

In this chapter we turn our attention to *double points* in $\mathbb{P}^1 \times \mathbb{P}^1$, that is, sets of fat points where every point has multiplicity two. Using our classification of ACM sets of fat points in $\mathbb{P}^1 \times \mathbb{P}^1$ that we developed in the previous chapter, it is not difficult to show that these sets of points are rarely ACM. In fact, even if the support of the set of fat points is ACM, it is not even true that the set of double points supported on this set of points is ACM as seen in Example 6.23.

However, even though these schemes may not be ACM, a set of double points in $\mathbb{P}^1 \times \mathbb{P}^1$ whose support is ACM still has a lot of structure that can be described combinatorially. In particular, if Z is a set of double points in $\mathbb{P}^1 \times \mathbb{P}^1$ whose support $X = \mathrm{Supp}(Z)$ is ACM, then we can describe the bigraded minimal free resolution of $I(Z)$ directly from the tuple α_X that describes the support.

The main result of this chapter, which is based upon [43] and [46], is an algorithm (Algorithm 7.24) whose input is the tuple α_X, and the output is the bigraded shifts that appear in the bigraded minimal free resolution of $I(Z)$, where Z is a collection of double points whose support is X.

This chapter, admittedly, is the most technical chapter in this monograph. The payoff comes in the next chapter when we exploit our bigraded minimal free resolution to apply our results to some problems about minimal free resolutions of ideals. To help guide the reader of the chapter, here are the broad strokes of the algorithm:

- Starting from Z, we construct a set Y of reduced and double points, which we call the *completion of* Z, such that $Z \subseteq Y$ and Y is ACM (see Theorem 7.4). We then use Theorem 6.27 to compute the bigraded minimal free resolution of $I(Y)$ from the tuple α_X associated with $X = \mathrm{Supp}(Z)$.
- Using [43] we use α_X to construct bihomogeneous forms $\{F_1, \ldots, F_p\}$ such that $I(Z) = I(Y) + (F_1, \ldots, F_p)$ and where $\deg F_i$ is a function of α_X (see Theorem 7.12).

© The Authors 2015

E. Guardo, A. Van Tuyl, *Arithmetically Cohen-Macaulay Sets of Points in* $\mathbb{P}^1 \times \mathbb{P}^1$,
SpringerBriefs in Mathematics, DOI 10.1007/978-3-319-24166-1_7

- For $j=0,\dots,p$, we set $I_0 = I(Y)$ and $I_j = (I_{j-1}, F_j)$. For each $j = 1,\dots,p$, we show (see Lemma 7.21) that $(I_{j-1} : F_j)$ is the defining ideal of a complete intersection of points whose type (and hence minimal resolution) can be computed from α_X.
- For each $j = 1,\dots,p$, we have a short exact sequence

$$0 \to R/(I_{j-1} : F_j)(-\deg F_j) \xrightarrow{\times \overline{F}_j} R/I_{j-1} \longrightarrow R/I_j \longrightarrow 0.$$

  We prove (cf. Theorem 7.22) that the mapping cone construction gives the bigraded minimal free resolution of R/I_j for each j.
- Because the minimal resolution of $I(Y) = I_0$ depends only upon α_X, we can iteratively use the mapping cone construction and the fact that $(I_{j-1} : F_j)$ is a complete intersection whose type can be computed from α_X to compute the minimal resolution $I(Z) = I_p$.

7.1 Setup

In this chapter, our goal is to describe the minimal bigraded free resolution of the ideals of the following special class of fat points in $\mathbb{P}^1 \times \mathbb{P}^1$.

Convention 7.1. For the remainder of this chapter, Z will denote a set of double points in $\mathbb{P}^1 \times \mathbb{P}^1$ with the property that $\mathrm{Supp}(Z) = X$ is an ACM set of points with $\alpha_X = (\alpha_1,\dots,\alpha_h)$.

The following example will be used as our running example for this chapter in order to illustrate the steps of our algorithm.

Example 7.2. Let X be as in Example 3.16. As shown in Example 4.12, this set of points is ACM with $\alpha_X = (6,5,3,1,1)$. Now consider the set of fat points Z defined by

$$I(Z) = \bigcap_{A_i \times B_j \in X} I(A_i \times B_j)^2.$$

This set of fat points is an example of a set of points that satisfies Convention 7.1. We can visualize this set of fat points as in Figure 7.1. In the figure, each dot represents a point, and the number its multiplicity. As we calculated in Example 6.23, for this set of fat points $Z \subseteq \mathbb{P}^1 \times \mathbb{P}^1$, we have

$$\alpha_Z = (12,10,6,6,5,3,2,2,1,1) \text{ and } \beta_Z = (10,6,6,5,4,4,3,3,2,2,2,1).$$

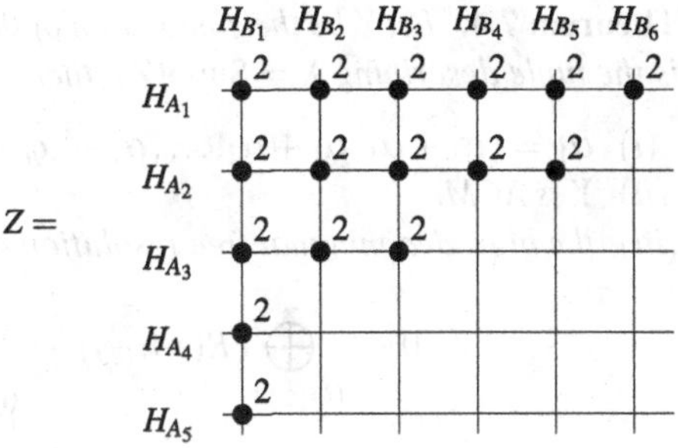

Fig. 7.1 Running example: double points whose support is ACM with $\alpha_X = (6,5,3,1,1)$

It follows that Z is not ACM because $\alpha_Z^* = (10,8,6,5,5,4,2,2,2,2,1,1) \neq \beta_Z$. Or, equivalently,

$$S_Z = \{(2,2,2,2,2,2),(2,2,2,2,2,0),(2,2,2,0,0,0),(2,0,0,0,0,0),(2,0,0,0,0,0),\\ (1,1,1,1,1,1),(1,1,1,1,1,0),(1,1,1,0,0,0),(1,0,0,0,0,0),(1,0,0,0,0,0)\}$$

has incomparable elements.

7.2 The completion of Z

Let Z be a set of double points that satisfies Convention 7.1, and let $\alpha_X = (\alpha_1,\ldots,\alpha_h)$ be the partition that describes the ACM support X. In this section we build a set of fat points Y, which we call the completion of Z, that contains Z. The set of fat points Y will be an ACM set of fat points. The bigraded minimal free resolution of $I(Y)$ will form the base step in our recursive formula to compute the bigraded resolution of $I(Z)$.

Geometrically, the completion of Z is formed by adding a number of simple (reduced) points to Z so that the support of the new set of fat points becomes a complete intersection. If X is the support of Z, and if $\pi_1(X) = \{A_1,\ldots,A_h\}$ and $\pi_2(X) = \{B_1,\ldots,B_v\}$, then

$$X \subseteq W = \{A_i \times B_j \mid A_i \in \pi_1(X) \text{ and } B_j \in \pi_2(X)\}.$$

Note that W is a complete intersection of reduced points.

Definition 7.3. Suppose that Z is a set of double points that satisfies Convention 7.1. With the notation as above, the *completion* of Z is the set of fat points $Y = Z \cup (W \backslash X)$.

The support of the completion is the complete intersection $CI(h,v)$. (Because of Convention 7.1, we have $v = \alpha_1$.) As we show below, we can use Theorem 6.21 to verify that Y is ACM. In fact, we can find the bigraded minimal free resolution of $I(Y)$ using Theorem 6.27. The key idea is to write α_Y in terms of α_X.

Theorem 7.4. *Let Y be the completion of the set of fat points Z. If $\alpha_X = (\alpha_1,\dots,\alpha_h)$ is the tuple describing $X = \mathrm{Supp}(Z)$, then*

(*i*) $\alpha_Y = (\alpha_1+\alpha_1, \alpha_1+\alpha_2, \dots, \alpha_1+\alpha_h, \alpha_1, \alpha_2, \dots, \alpha_h)$.
(*ii*) *Y is ACM.*
(*iii*) *the bigraded minimal free resolution of $I(Y)$ has the form*

$$0 \to \bigoplus_{(i,j)\in SY_1} R(-i,-j) \to \bigoplus_{(i,j)\in SY_0} R(-i,-j) \to I(Y) \to 0$$

where

$$\begin{aligned} SY_0 &= \{(2h,0),(h,\alpha_1),(0,2\alpha_1)\} \\ &\quad \cup \{(i-1,\alpha_1+\alpha_i)\ (i+h-1,\alpha_i) \mid \alpha_i - \alpha_{i-1} < 0\} \\ SY_1 &= \{(2h,\alpha_h),(h,\alpha_1+\alpha_h)\} \\ &\quad \cup \{(i-1,\alpha_1+\alpha_{i-1}),(i+h-1,\alpha_{i-1}) \mid \alpha_i - \alpha_{i-1} < 0\}. \end{aligned}$$

Proof. Statement (*i*) follows directly from the construction of Y. For statement (*ii*) it suffices to note that if $\alpha_X^* = (\alpha_1^*,\dots,\alpha_{\alpha_1}^*)$, then $\beta_Y = (\alpha_1^*+\alpha_1^*,\dots,\alpha_1^*+\alpha_{\alpha_1}^*, \alpha_1^*,\dots,\alpha_{\alpha_1}^*)$. Moreover, one can check that $\alpha_Y^* = \beta_Y$, so that by Theorem 6.21 it follows that Y is ACM. The bigraded resolution of (*iii*) is an application of Theorem 6.27.

Remark 7.5. Unlike Theorems 5.3 and 6.21 where we used C_W and V_W to denote the set of bigraded shifts in the minimal free resolution of a set of fat points W, we will instead use SW_i to denote the bigraded shifts in homological degree i in the bigraded minimal free resolution of $I(W)$. The change in our notation is required because our set of fat points may not be ACM, so our resolutions may be longer than those of Theorems 5.3 and 6.21.

Example 7.6. Let Z be the set of fat points of Example 7.2. We can visualize the completion of Z as in Figure 7.2 where • means a double point and ∘ means a simple point (we have suppressed the multiplicities). Because $\alpha_Z = (6,5,3,1,1)$,

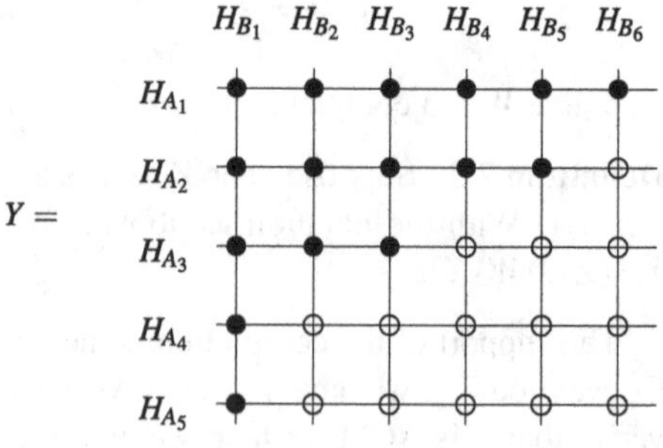

Fig. 7.2 The completion of Z

we have $\alpha_Y = (12,11,9,7,7,6,5,3,1,1)$. Then the shifts in the bigraded minimal free resolution of $I(Y)$ are given by

$$SY_0 = \{(10,0),(8,1),(7,3),(6,5),(5,6),(3,7),(2,9),(1,11),(0,12)\}$$
$$SY_1 = \{(10,1),(8,3),(7,5),(6,6),(5,7),(3,9),(2,11),(1,12)\}.$$

7.3 The generators of $I(Z)$ and $I(Y)$

In this section, we use the tuple α_X to construct a matrix whose entries are either two or one. We then extract information from this matrix to describe the minimal generators of $I(Z)$ and $I(Y)$. Specifically, because $I(Y) \subseteq I(Z)$, we will identify a family of bigraded forms $\{F_1,\dots,F_p\}$ such that $F_i \notin I(Y)+(F_1,\dots,F_{i-1})$ for $i=1,\dots,p$, and $I(Z) = I(Y)+(F_1,\dots,F_p)$.

Definition 7.7. If $\alpha = \alpha_X = (\alpha_1,\dots,\alpha_h)$ is the partition associated with $X = \mathrm{Supp}(Z)$, then the *degree matrix* of Z is the $h \times \alpha_1$ matrix M_α where

$$(M_\alpha)_{i,j} = \begin{cases} 2 & j \leq \alpha_i \\ 1 & \text{otherwise.} \end{cases}$$

Remark 7.8. If the points in the support of Z have been relabeled according to Convention 3.9, then $(M_\alpha)_{i,j}$ is the multiplicity of the point $A_i \times B_j$ in Y, the completion of Z.

Using the degree matrix of Z, we identify certain positions in the matrix.

Definition 7.9. The *base corners* of Z is the set:

$$C_0 := \{(i,j) \mid (M_\alpha)_{i,j} = 1 \text{ but } (M_\alpha)_{i-1,j} = (M_\alpha)_{i,j-1} = 2\}.$$

Given the base corners of Z, we then set

$$C_1 := \{(i,l) \mid (i,j),(k,l) \in C_0 \text{ and } i > k\}.$$

The *corners* of Z is then the set $C := C_0 \cup C_1$. We shall assume that the elements of C have been ordered from largest to smallest with respect to the lexicographical order.

Remark 7.10. The set of base corners C_0 can be computed directly from the partition α_X associated with $\mathrm{Supp}(Z)$. Precisely, $C_0 := \{(i,\alpha_i+1) \mid \alpha_i - \alpha_{i-1} < 0\}$.

Definition 7.11. For each $(i,j) \in C$, set

$$u_{i,j} := m_{1,j} + m_{2,j} + \cdots + m_{i-1,j} \text{ and } v_{i,j} := m_{i,1} + m_{i,2} + \cdots + m_{i,j-1}$$

were $m_{a,b} = (M_\alpha)_{a,b}$. That is, $u_{i,j}$, respectively $v_{i,j}$, is the sum of the entries in M_α in the column above, respectively in the row to the left, of the position (i,j). If $(i,j) = (i_\ell, j_\ell)$ is the ℓ-th largest element of C with respect to the lexicographical order, the form

$$F_\ell = H_{A_1}^{m_{1,j}} \cdots H_{A_{i-1}}^{m_{i-1,j}} V_{B_1}^{m_{i,1}} \cdots V_{B_{j-1}}^{m_{i,j-1}} \quad \text{where } m_{a,b} = (M_\alpha)_{a,b}$$

is called the *form relative to the corner* (i,j).

Theorem 7.12. *Let Z be a set of fat points that satisfies Convention 7.1, and furthermore, assume that the points in the support have been relabeled using Convention 3.9. If $(i,j) = (i_\ell, j_\ell)$ is the ℓ-th largest element of C with respect to the lexicographical order, then let*

$$F_\ell = H_{A_1}^{m_{1,j}} \cdots H_{A_{i-1}}^{m_{i-1,j}} V_{B_1}^{m_{i,1}} \cdots V_{B_{j-1}}^{m_{i,j-1}}$$

be the form relative to the corner (i,j). Set $I_0 := I(Y)$, and $I_\ell := (I_{\ell-1}, F_\ell)$ for $\ell = 0, \ldots, |C|$. Then

(*i*) $\deg F_\ell = (u_{i,j}, v_{i,j})$.
(*ii*) $F_\ell \notin I_{\ell-1}$.
(*iii*) $I(Z) = I(Y) + (F_1, \ldots, F_p)$ *where* $p = |C|$.
(*iv*) *I_ℓ is generated by the generators of $I(Y)$, and all the forms relative to corners (a,b) with (a,b) bigger than or equal to (i_ℓ, j_ℓ).*

Proof. Statement (i) is immediate from the definition of F_ℓ. For statement (ii), note that after relabeling, $P_{i_\ell, j_\ell} = A_{i_\ell} \times B_{j_\ell}$ is a reduced point of Y. Furthermore, every element of $I_{\ell-1}$ vanishes at the point P_{i_ℓ, j_ℓ}, i.e., $I_{\ell-1} \subseteq I_{P_{i_\ell, j_\ell}} = (H_{A_{i_\ell}}, V_{B_{j_\ell}})$, but the form $F_\ell \notin I_{P_{i_\ell, j_\ell}}$. Statements (iii) and (iv) are Theorem 3.15 of [43].

Remark 7.13. Although it was not said explicitly, we can observe that each F_ℓ is a minimal separator of the point P_{i_ℓ, j_ℓ} (see Definition 3.34) in the set of points defined by the ideal $I_{\ell-1}$.

A slight variation of the above technique enables us to describe the generators of $I(Y)$.

Definition 7.14. Let $\alpha_X = (\alpha_1, \ldots, \alpha_h)$ be the partition associated with Supp(Z), and suppose M_α is the *degree matrix* of Z. The *degree matrix* of Y is the $(h+1) \times (\alpha_1 + 1)$ matrix

$$M_Y = \begin{bmatrix} M_\alpha & \mathbf{1} \\ \mathbf{1} & 1 \end{bmatrix}$$

where **1** denotes the appropriately sized matrix consisting only of ones.

Definition 7.15. Let C_0 be the base corners of Z constructed from $\alpha_X = (\alpha_1,\dots,\alpha_h)$. The *outside corners* of Z is the set

$$OC = \{(h+1,1),(1,\alpha_1+1),(h+1,\alpha_1+1)\} \cup \{(h+1,j),(i,\alpha_1+1) \mid (i,j) \in C_0\}.$$

Theorem 7.16. *Let Z be a set of fat points that satisfies Convention 7.1, and furthermore, assume that the points in the support have been relabeled using Convention 3.9. If $(i,j) = (i_\ell, j_\ell) \in OC$, then set*

$$G_\ell = H_{A_1}^{m_{1,j}} \cdots H_{A_{i-1}}^{m_{i-1,j}} V_{B_1}^{m_{i,1}} \cdots V_{B_{j-1}}^{m_{i,j-1}} \text{ where } m_{a,b} = (M_Y)_{a,b}.$$

Then $\{G_1,\dots,G_q\}$ where $q = |OC|$ is a minimal set of generators of $I(Y)$.

Proof. For each $\ell = 1,\dots,q$, one can show that G_ℓ passes through all the points of Y to the correct multiplicity. By comparing the degrees of each G_ℓ with the degrees of the minimal generators of $I(Y)$ from the bigraded minimal free resolution in Theorem 7.4, we then see that the G_ℓ's form a minimal set of generators of $I(Y)$.

We end this section with an example illustrating these ideas.

Example 7.17. Let $\alpha_X = (6,5,3,1,1)$ be the tuple associated with the support of the set of fat points Z of Example 7.2. Then the degree matrices of Z and Y are given by

$$M_\alpha = \begin{bmatrix} 2&2&2&2&2&2 \\ 2&2&2&2&2&\underline{1} \\ 2&2&2&\underline{1}&1&\underline{1} \\ 2&\underline{1}&1&\underline{1}&1&\underline{1} \\ 2&1&1&1&1&1 \end{bmatrix} \quad M_Y = \begin{bmatrix} 2&2&2&2&2&2&\underline{1} \\ 2&2&2&2&2&1&\underline{1} \\ 2&2&2&1&1&1&\underline{1} \\ 2&1&1&1&1&1&\underline{1} \\ 2&1&1&1&1&1&1 \\ \underline{1}&\underline{1}&1&\underline{1}&1&\underline{1}&\underline{1} \end{bmatrix}.$$

Then $C_0 = \{(4,2),(3,4),(2,6)\}$, ordered lexicographically. The set of the corners of Z is

$$C := C_0 \cup \{(4,4),(4,6),(3,6)\} = \{(4,6),(4,4),(4,2),(3,6),(3,4),(2,6)\}.$$

The positions of the underlined 1's in M_α correspond to the elements of C.

The outside corners, which correspond to the positions of the underlined 1's in the matrix M_Y, is the set

$$OC = \{(6,1),(6,2),(6,4),(6,6),(6,7),(1,7),(2,7),(3,7),(4,7)\}.$$

As an example of Theorem 7.16, consider $(6,6) \in OC$. Associated with this tuple is the form

$$G = H_{A_1}^2 H_{A_2}^1 H_{A_3}^1 H_{A_4}^1 H_{A_5}^1 V_{B_1}^1 V_{B_2}^1 V_{B_3}^1 V_{B_4}^1 V_{B_5}^1.$$

We see from the picture of Example 7.6 that G passes through all the points (with correct multiplicity) of Y. Also, $\deg G = (6,5)$ is one of the degrees of the minimal generators.

Observation 7.18. The following fact will be used implicitly in the next section. For each $(i,j) \in C$ there exist non-negative integers c and d such that $(i+c+1,j)$, $(i,j+d+1)$ and $(i+c+1,j+d+1)$ are either elements of C or OC. Although we leave the proof of this fact to the reader, we can illustrate this observation using the above example. Note that $(4,2)$ is a corner of Z. There exist two integers $c=1$ and $d=1$ such that $(4+1+1,2)$, $(4,2+1+1)$ and $(4+1+1,2+1+1)$ are also corners or outside corners.

7.4 The resolution of $I(Z)$

Let $F_1,\ldots,F_p$ be the p forms of Theorem 7.12 where F_ℓ is the form relative to the corner $(i_\ell,j_\ell) \in C$. As in Theorem 7.12, we set $I_0 = I(Y)$ and $I_\ell = (I_{\ell-1},F_\ell)$ for $\ell = 1,\ldots,p$. Then, for each $1 \leq \ell \leq p$, we have a short exact sequence

$$0 \to R/(I_{\ell-1}:F_\ell)(-u_{i_\ell,j_\ell},-v_{i_\ell,j_\ell}) \xrightarrow{\times F_\ell} R/I_{\ell-1} \to R/I_\ell = R/(I_{\ell-1},F_\ell) \to 0 \quad (7.1)$$

where $\deg F_\ell = (u_{i_\ell,j_\ell},v_{i_\ell,j_\ell})$. Using the short exact sequence and the mapping cone construction, we will iteratively describe the bigraded minimal free resolution of $I(Z)$.

To use the mapping cone construction in conjunction with (7.1), we will prove that $(I_{\ell-1}:F_\ell)$ is a complete intersection for each $\ell = 1,\ldots,p$ whose type can be determined through the following family of matrices. Let $C = \{(i_1,j_1),\ldots,(i_p,j_p)\}$ be the corners of Z ordered from largest to smallest with respect to the lexicographical order. Then set $M_0 = M_\alpha$, and for $\ell = 1,\ldots,p$, let M_ℓ be the $h \times \alpha_1$ matrix where

$$(M_\ell)_{i,j} = \begin{cases} 0 & \text{if } (i,j) \succeq (i_\ell,j_\ell) \\ (M_{\ell-1})_{i,j} & \text{otherwise.} \end{cases}$$

Here $\succeq$ denotes the partial order where $(i_1,j_1) \succeq (i_2,j_2)$ if and only if $i_1 \geq i_2$ and $j_1 \geq j_2$.

Example 7.19. Before proceeding to the main results of this paper, we describe in more detail what our algorithm does geometrically, and how we shall use the matrices M_ℓ. Let Z_ℓ denote the set of fat points defined by the ideal I_ℓ, where $Z_0 = Y$ is the completion of Z. Roughly speaking, at each step in our algorithm, we are removing a set of points from $Z_{\ell-1}$ to form the set of points Z_ℓ. In particular, at each step we are removing a complete intersection whose type can be ascertained from the matrix $M_{\ell-1}$.

We illustrate some of these ideas by using our running example (Example 7.2) with $\alpha = (6,5,3,1,1)$. The matrix $M_0 = M_\alpha$ of Example 7.17 describes the multiplicities of the fat points $Z_0 = Y$. By Example 7.17 the largest corner of Z is $(4,6)$. The element

$$F_1 = H_{A_1}^2 H_{A_2} H_{A_3} V_{B_1}^2 V_{B_2} V_{B_3} V_{B_4} V_{B_5}$$

is the form relative to the corner $(4,6)$. The form F_1 passes through all the points of $Z_0 = Y$ with correct multiplicity, except the points $A_a \times B_b$ with $(4,6) \preceq (a,b) \preceq (5,6)$. These points are $C = \{A_4 \times B_6, A_5 \times B_6\}$, a complete intersection of points of type $(2,1)$ defined by $I_C = (H_{A_4} H_{A_5}, V_{B_6})$. The type can be found by starting at the location of the first corner $(4,6)$ in M_0, and summing the entry in position $(4,6)$ and all those below it (in this case, $1+1=2$), to get the first coordinate of the type, and summing the entry in position $(4,6)$ and all those to right (in this case, only 1) to get the second coordinate of the type.

The ideal $I_1 = (I_0, F_1)$ is then the defining ideal of Z_1, where

$$Z_1 = Y \backslash CI(2,1) = Y \backslash \{A_4 \times B_6, A_5 \times B_6\}.$$

Observe now that the matrix

$$M_1 = \begin{bmatrix} 2&2&2&2&2&2 \\ 2&2&2&2&2&1 \\ 2&2&2&1&1&1 \\ 2&1&1&1&1&0 \\ 2&1&1&1&1&0 \end{bmatrix}$$

describes the multiplicities of the set of fat points Z_1 as seen in Figure 7.3, where $\bullet$ means a double point and $\circ$ means a simple point.

The next largest corner of Z is $(4,4)$, and the form

$$F_2 = H_{A_1}^2 H_{A_2}^2 H_{A_3} V_{B_1}^2 V_{B_2} V_{B_3}$$

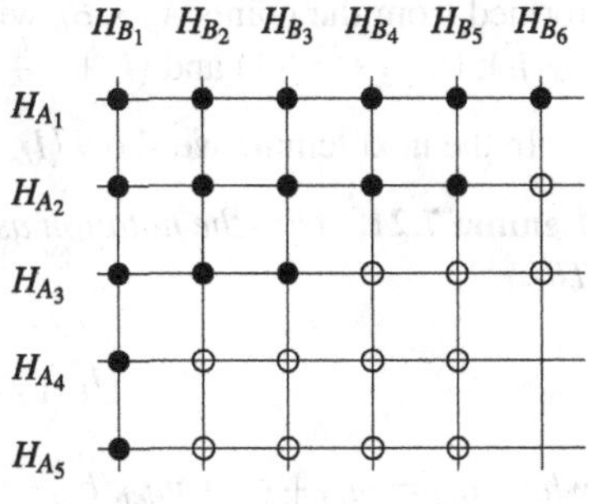

Fig. 7.3 The set of fat points Z_1

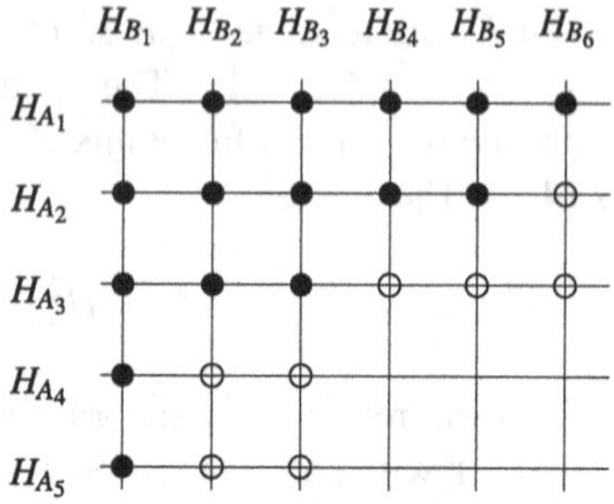

Fig. 7.4 The set of fat points Z_2

is the form relative to the second corner $(4,4)$. The form F_2 now passes through all the points of the set of fat points Z_1 with correct multiplicity, except the points $A_a \times B_b$ with $(4,4) \preceq (a,b) \preceq (5,5)$. These points are $C = \{A_4 \times B_4, A_4 \times B_5, A_5 \times B_4, A_5 \times B_5\}$, a complete intersection of type $(2,2)$ defined by $I_C = (H_{A_4}H_{A_5}, V_{B_4}V_{B_5})$. The type can be found by starting at the location of the second corner $(4,4)$ in M_1, and summing the entry in position $(4,4)$ and all those below it (in this case, $1+1=2$), to get the first coordinate of the type, and summing the entry in position $(4,4)$ and all those to right (in this case, $1+1+0=2$) to get the second coordinate.

The ideal $I_2 = (I_1, F_1)$ now defines the set of fat points

$$\begin{aligned} Z_2 &= Z_1 \backslash CI(2,2) = Z_1 \backslash \{A_4 \times B_4, A_4 \times B_5, A_5 \times B_4, A_5 \times B_5\} \\ &= Y \backslash \{A_4 \times B_4, A_4 \times B_5, A_4 \times B_6, A_5 \times B_4, A_5 \times B_5, A_5 \times B_6\} \end{aligned}$$

and analogously, the matrix M_2 describes the multiplicities of the set of fat points Z_2 as shown in Figure 7.4.

Continuing in this fashion, we remove all the simple points from Y by removing a suitably sized complete intersection at each step until we get $Z_6 = Z$. In general, the matrices M_ℓ allow us to keep track of the size of the complete intersection we are cutting out from Z_ℓ at each step.

Remark 7.20. Let $\{(i_1,j_1),\ldots,(i_p,j_p)\}$ be the corners of Z starting from the largest corner of Z. The complete intersection C that we remove at each step from Y is formed from the points $A_a \times B_b$ with $(i_\ell,j_\ell) \preceq (a,b) \preceq (i_\ell+c, j_\ell+d)$ and such that (i_ℓ,j_ℓ), $(i_\ell, j_\ell+c+1)$ and $(i_\ell+d+1, j_\ell)$ are either corners or outside corners of Z.

In the next lemma we show $(I_{\ell-1} : F_\ell)$ is a complete intersection of points.

Lemma 7.21. *With the notation as above, let $(i,j) = (i_\ell, j_\ell)$ be the ℓ-th corner of C. Then*

$$(I_{\ell-1} : F_\ell) = I(CI(a_{i,j}, b_{i,j}))$$

where $a_{i,j} = m_{i,j} + \cdots + m_{h,j}$, $b_{i,j} = m_{i,j} + \cdots + m_{i,\alpha_1}$ and $m_{a,b} = (M_{\ell-1})_{a,b}$.

Proof. Without loss of generality, assume that the points of Z have been relabeled in accordance with Convention 3.9. From the construction of $M_{\ell-1}$ there exist integers c and d such that $m_{i,j} = m_{i+1,j} = \cdots = m_{i+c,j} = 1$, but $m_{i+c+1,j} = \cdots = m_{h,j} = 0$, and similarly, $m_{i,j} = \cdots = m_{i,j+d} = 1$, but $m_{i,j+d+1} = \cdots = m_{i,\alpha_1} = 0$. Set

$$A = H_{A_i}^{m_{i,j}} \cdots H_{A_{i+c}}^{m_{i+c,j}} = H_{A_i} \cdots H_{A_{i+c}} \text{ and } B = V_{B_j}^{m_{i,j}} \cdots V_{B_{j+d}}^{m_{i,j+d}} = V_{B_j} \cdots V_{B_{j+d}}.$$

It will now suffice to show that $(I_{\ell-1} : F_\ell) = (A,B)$.

Note that (A,B) defines a complete intersection $C = CI(a_{i,j}, b_{i,j})$. Because the points have been rearranged in accordance with Convention 3.9, $A_a \times B_b \in C$ if and only if $(i,j) \preceq (a,b) \preceq (i+c, j+d)$. The points of C form a subset of the reduced points of Y.

By Theorems 7.12 and 7.16, $I_{\ell-1} = (G_1, \ldots, G_q, F_1, \ldots, F_{\ell-1})$. The forms G_i vanish at all the points of $C \subseteq Y$. By Theorem 7.12 we have $F_i \in I_C$ for $1 \leq i \leq \ell-1$. However,

$$F_\ell = H_{A_1}^{m_{1,j}} \cdots H_{A_{i-1}}^{m_{i-1,j}} V_{B_1}^{m_{i,1}} \cdots V_{B_{j-1}}^{m_{i,j-1}}$$

from which it follows that for every $A_a \times B_b \in C$, $F_\ell(A_a \times B_b) \neq 0$. So, if $HF_\ell \in I_{\ell-1} \subseteq I_C$, then $H \in I_C$. That is, $(I_{\ell-1} : F_\ell) \subseteq I_C = (A,B)$.

From the construction of $M_{\ell-1}$, $(i+c+1, j)$ is either a corner or outside corner of Z. In either case, set

$$F = H_{A_1}^{n_{1,j}} \cdots H_{A_{i-1}}^{n_{i-1,j}} H_{A_i}^{n_{i,j}} \cdots H_{A_{i+c}}^{n_{i+c,j}} V_{B_1}^{n_{i+c+1,1}} \cdots V_{B_{j-1}}^{n_{i+c+1,j-1}}$$

where $n_{a,b}$ refers to the entries in $M_Y = (n_{a,b})$, the degree matrix of Y. If $(i+c+1, j) \in C$, then $F \in I_{\ell-1}$ by Theorem 7.12; if $(i+c+1, j) \in OC$, then $F \in I_{\ell-1}$ by Theorem 7.16. Now set

$$F_\ell A = H_{A_1}^{m_{1,j}} \cdots H_{A_{i-1}}^{m_{i-1,j}} H_{A_i} \cdots H_{A_{i+c}} V_{B_1}^{m_{i,1}} \cdots V_{B_{j-1}}^{m_{i,j-1}}$$

We claim that F divides $F_\ell A$, and hence $F_\ell A \in I_{\ell-1}$. To see this we compare the matrices M_Y and $M_{\ell-1}$. By construction $(M_Y)_{a,b} = (M_\alpha)_{a,b} = (M_{\ell-1})_{a,b}$ for all $(a,b) \preceq (i+c, j)$. So, the exponents of the H_{A_i}'s in $F_\ell A$ and F are actually the same.

On the other hand, note that $n_{a,j} \geq n_{b,j}$ if $a \geq b$ in M_Y, i.e., the columns are non-increasing. Since $m_{i,t} = n_{i,t}$ for $t = 1, \ldots, j-1$, we have that the exponents of the V_{B_j}'s in F are less than or equal than those that appear in $F_\ell A$. So, F divides $F_\ell A$. So $A \in (I_{\ell-1} : F_\ell)$. A similar argument using the fact that $(i, j+d+1) \in C$ or OC will now show that $B \in (I_{\ell-1} : F_\ell)$. Hence $(A,B) \subseteq (I_{\ell-1} : F_\ell)$.

We now come to the main result of this section, which forms the basis of our recursive algorithm to compute the resolution of $I(Z)$.

Theorem 7.22. *With the notation as above, suppose that* $(i,j) = (i_\ell, j_\ell)$ *is the* ℓ*-th largest element of* C*, and furthermore, suppose that*

$$0 \to \mathbb{F}_2 \to \mathbb{F}_1 \to \mathbb{F}_0 \to I_{\ell-1} \to 0$$

is the bigraded minimal free resolution of $I_{\ell-1}$. *Then*

$$0 \to \begin{matrix} \mathbb{F}_2 \\ \oplus \\ R(-u_{i,j}-a_{i,j},-v_{i,j}-b_{i,j}) \end{matrix} \to \begin{matrix} \mathbb{F}_1 \\ \oplus \\ R(-u_{i,j}-a_{i,j},-v_{i,j}) \\ \oplus \\ R(-u_{i,j},-v_{i,j}-b_{i,j}) \end{matrix} \to \begin{matrix} \mathbb{F}_0 \\ \oplus \\ R(-u_{i,j},-v_{i,j}) \end{matrix} \to I_\ell \to 0 \tag{7.2}$$

is a bigraded minimal free resolution of $I_\ell = (I_{\ell-1}, F_\ell)$ *where*

$$u_{i,j} = m_{1,j} + m_{2,j} + \cdots + m_{i-1,j} \quad \textit{and} \quad v_{i,j} = m_{i,1} + m_{i,2} + \cdots + m_{i,j-1}$$

$$a_{i,j} = m_{i,j} + \cdots + m_{h,j} \quad \textit{and} \quad b_{i,j} = m_{i,j} + \cdots + m_{i,\alpha_1}$$

and $m_{a,b} = (M_{\ell-1})_{a,b}$.

Proof. Let $(i,j) = (i_\ell, j_\ell) \in C$ denote the ℓ-th largest corner of Z, and assume that the points of Z have been rearranged in accordance with Convention 3.9. Let

$$F_\ell = H_{A_1}^{m_{1,j}} \cdots H_{A_{i-1}}^{m_{i-1,j}} V_{B_1}^{m_{i,1}} \cdots V_{B_{j-1}}^{m_{i,j-1}}$$

be the form relative to the corner (i,j) with $\deg F_\ell = (u_{i,j}, v_{i,j})$. Note that for all (a,b) with $(a,b) \preceq (i,j)$, we have $(M_{\ell-1})_{a,b} = (M_\alpha)_{a,b}$. So, the integers $u_{i,j}$ and $v_{i,j}$ as defined above are the same as those of Theorem 7.12.

By Lemma 7.21, we know that $(I_{\ell-1} : F_\ell) = I(CI(a_{i,j}, b_{i,j}))$. Because this ideal is a complete intersection, a bigraded minimal free resolution of $(I_{\ell-1} : F_\ell)$ is given by Lemma 2.26:

$$0 \to R(-a_{i,j}, -b_{i,j}) \to R(-a_{i,j}, 0) \oplus R(0, -b_{i,j}) \to (I_{\ell-1} : F_\ell) \to 0.$$

When we apply the mapping cone construction to the short exact sequence (7.1), we get that (7.2) is a bigraded free resolution of I_ℓ. It therefore suffices to verify that this resolution is minimal.

The map in (7.1)

$$R/(I_{\ell-1} : F_\ell)(-u_{i,j}, -v_{i,j}) \xrightarrow{\times F_\ell} R/I_{\ell-1}$$

lifts to a map from the minimal resolution of $R/(I_{\ell-1} : F_\ell)$ to that of $R/I_{\ell-1}$:

$$\begin{array}{ccccccccc} 0 \to R & \xrightarrow{\phi_1} & R^2 & \xrightarrow{\phi_0} & R & \xrightarrow{\varepsilon} & R/(I_{\ell-1} : F_\ell) & \to 0 \\ \downarrow \delta_2 & & \downarrow \delta_1 & & \downarrow \times F_\ell & & \downarrow \times F_\ell & \\ 0 \to \mathbb{F}_2 \xrightarrow{\varphi_2} \mathbb{F}_1 & \xrightarrow{\varphi_1} & \mathbb{F}_0 & \xrightarrow{\varphi_0} & R & \xrightarrow{\varepsilon} & R/I_{\ell-1} & \to 0. \end{array}$$

We have suppressed all the shifts in the resolutions. The maps in each square commute. Again suppressing the shifts, the resolution of R/I_ℓ given by the mapping cone construction has the form

$$0 \to R \oplus \mathbb{F}_2 \xrightarrow{\Phi_2} R^2 \oplus \mathbb{F}_1 \xrightarrow{\Phi_1} R \oplus \mathbb{F}_0 \xrightarrow{\Phi_0} R \to R/I_\ell \to 0$$

where the maps are

$$\Phi_2 = \begin{bmatrix} -\phi_1 & 0 \\ \delta_2 & \varphi_2 \end{bmatrix}, \quad \Phi_1 = \begin{bmatrix} -\phi_0 & 0 \\ \delta_1 & \varphi_1 \end{bmatrix}, \quad \text{and } \Phi_0 = \begin{bmatrix} F_\ell & \varphi_0 \end{bmatrix}.$$

After fixing a basis, each map ϕ_i, φ_i, and δ_i can be represented by a matrix with entries in S. It will therefore suffice to show that all the nonzero entries of the matrix corresponding to the map Φ_i for $i = 0,1,2$ belong to the maximal ideal (x_0,x_1,y_0,y_1) of R. The matrices corresponding to ϕ_i and φ_i already have this property because they are the maps in the minimal resolution of $R/(I_{\ell-1} : F_\ell)$ and $R/I_{\ell-1}$, respectively. So, we need to show that there exist maps δ_1 and δ_2 that make each square commute, and when these maps are represented as matrices, all the nonzero entries belong to (x_0,x_1,y_0,y_1).

From Observation 7.18, because $(i,j) \in C$, there exist integers c and d such that $(i+c+1,j)$, $(i,j+d+1)$, and $(i+c+1,j+d+1)$ are either corners or outside corners of Z; in particular, we choose c and d as in the proof of Lemma 7.21, that is, $m_{i,j} = m_{i+1,j} = \cdots = m_{i+c,j} = 1$, but $m_{i+c+1,j} = \cdots = m_{r,j} = 0$, and similarly, $m_{i,j} = \cdots = m_{i,j+d} = 1$, but $m_{i,j+d+1} = \cdots = m_{i,\alpha_1} = 0$ with $m_{a,b} = (M_{\ell-1})_{a,b}$. Set

$$A = H_{A_i}^{m_{i,j}} \cdots H_{A_{i+c}}^{m_{i+c,j}} = H_{A_i} \cdots H_{A_{i+c}} \text{ and } B = V_{B_j}^{m_{i,j}} \cdots V_{B_{j+d}}^{m_{i,j+d}} = V_{B_j} \cdots V_{B_{j+d}}.$$

Because $(I_{\ell-1} : F_\ell) = (A,B)$ is a complete intersection, the maps ϕ_0 and ϕ_1 are simply the Koszul maps. As matrices, these maps are

$$\phi_1 = \begin{bmatrix} B \\ -A \end{bmatrix} \text{ and } \phi_0 = \begin{bmatrix} A & B \end{bmatrix}.$$

We also let

$$H_1 = H_{A_1}^{n_{1,j}} \cdots H_{A_{i+c}}^{n_{i+c,j}} V_{B_1}^{n_{i+c+1,1}} \cdots V_{B_{j-1}}^{n_{i+c+1,j-1}}$$

$$H_2 = H_{A_1}^{n_{1,j+d+1}} \cdots H_{A_{i-1}}^{n_{i-1,j+d+1}} V_{B_1}^{n_{i,1}} \cdots V_{B_{j+d}}^{n_{i,j+d}}$$

$$H_3 = H_{A_1}^{n_{1,j+d+1}} \cdots H_{A_{i+c}}^{n_{i+c,j+d+1}} V_{B_1}^{n_{i+c+1,1}} \cdots V_{B_{j+d}}^{n_{i+c+1,j+d}}.$$

where $n_{a,b} = (M_Y)_{a,b}$.

Now $(i+c+1,j)$, $(i,j+d+1)$, and $(i+c+1,j+d+1)$ are either corners or outside corners of Z. In the case that they are corners of Z, then they are larger than the corner (i,j). So by Theorems 7.12 and 7.16 we have that the forms H_1,H_2,H_3 are minimal generators of $I_{\ell-1}$.

After a suitable change of basis, we can then write φ_0 as

$$\varphi_0 = \begin{bmatrix} H_1 & H_2 & H_3 & K_1 & \cdots & K_s \end{bmatrix}$$

where $K_1,\ldots,K_s$ denote the other minimal generators of $I_{\ell-1}$.

Let

$$C = \frac{F_\ell A}{H_1} = \frac{H_{A_1}^{m_{1,j}} \cdots H_{A_{i-1}}^{m_{i-1,j}} V_{B_1}^{m_{i,1}} \cdots V_{B_{j-1}}^{m_{i,j-1}} H_{A_i} \cdots H_{A_{i+c}}}{H_{A_1}^{n_{1,j}} \cdots H_{A_{i+c}}^{n_{i+c,j}} V_{B_1}^{n_{i+c+1,1}} \cdots V_{B_{j-1}}^{n_{i+c+1,j-1}}}.$$

Now, by the construction of M_Y and $M_{\ell-1}$, we also have $(M_Y)_{a,b} = (M_{\ell-1})_{a,b}$ for all $(a,b) \preceq (i+c,j+d)$. The exponents of the H_{A_i}'s in the above expression are then the same on the top and bottom, and thus they cancel out, i.e.,

$$C = \frac{F_\ell A}{H_1} = \frac{V_{B_1}^{m_{i,1}} \cdots V_{B_{j-1}}^{m_{i,j-1}}}{V_{B_1}^{n_{i+c+1,1}} \cdots V_{B_{j-1}}^{n_{i+c+1,j-1}}}.$$

Because (i,j) is a corner and $(i+c+1,j)$ is either a corner or outside corner of Z, by the construction of the matrix M_Y, there exists some $j' \leq j-1$ such that $n_{i+c+1,j'} < n_{i,j'} = m_{i,j'}$. (The columns of M_Y are non-increasing, so if $n_{i+c+1,j'} = n_{i,j'}$ for all $j' \leq j-1$, then the first $j-1$ entries of rows i through $i+c+1$ are the same, and thus there would not be a corner (or outside corner) in position $(i+c+1,j)$.) Because of this fact, we have $\deg C > 0$. A similar argument implies that if $D = \frac{F_\ell B}{H_2}$, then $\deg D > 0$.

Because $F_\ell H_3 = H_1 H_2$, we have the following two syzygies:

$$BH_1 - DH_3 = 0 \text{ and } AH_2 - CH_3 = 0.$$

That is, $(B,0,-D,0,\ldots,0)^T$ and $(0,A,-C,0,\ldots,0)^T$ are two elements of $\mathbb{F}_0$, written as vectors, in $\ker \varphi_0 = \text{Im } \varphi_1$. Let $\underline{a} = (a_1,\ldots,a_m)^T$, respectively, let $\underline{b} = (b_1,\ldots,b_m)^T$ denote an element of $\mathbb{F}_1$ with $\varphi_1(\underline{a}) = (B,0,-D,0,\ldots,0)^T$, respectively, $\varphi_1(\underline{b}) = (0,A,-C,0,\ldots,0)^T$. With this notation, we can now prove:

Claim. The maps δ_1 and δ_2 are given by

$$\delta_2 = \begin{bmatrix} Ca_1 - Db_1 \\ \vdots \\ Ca_m - Db_m \end{bmatrix} \quad \text{and} \quad \delta_1 = \begin{bmatrix} C & 0 \\ 0 & D \\ 0 & 0 \\ \vdots & \vdots \\ 0 & 0 \end{bmatrix}.$$

Proof. We just need to show that each square containing a δ_i commutes. Now $\varphi_0\delta_1 = \left[H_1C\ H_2D\right] = \left[F_\ell A\ F_\ell B\right]$. This map is the same as composing the map ϕ_0 with the map defined by multiplication by F_ℓ. For the second square,

$$\begin{aligned}\varphi_1\delta_2 &= C\varphi_1(\underline{a}) - D\varphi_1(\underline{b}) = C(B,0,-D,0,\ldots,0)^T - D(0,A,-C,0,\ldots,0)^T \\ &= (CB,-DA,0,\ldots,0)^T = \delta_1\phi_1.\end{aligned}$$

This completes the proof of the claim.

Because C and D are non-constant bihomogeneous forms, every nonzero entry of δ_1 and δ_2 belongs to $(x_0,x_1,y_0,y_1) \subseteq R$. Therefore, the resolution of I_ℓ is minimal, as desired.

Remark 7.23. As observed in Example 7.19, the ideal I_ℓ corresponds to a subset of Y formed by removing a number of complete intersections of reduced points. The above theorem allows us to calculate the bigraded minimal free resolution for each such subset "between" Y and Z, that is, those sets of fat points we called Z_ℓ in Example 7.19.

7.5 The algorithm

The resolution of the completion $I_0 = I(Y)$ depends only upon $\alpha = \alpha_X$. By repeatedly applying Theorem 7.22, we obtain the minimal resolution of $I_p = I(Z)$. Furthermore, the shifts that appear at each step only depend upon $M_{\ell-1}$ which is constructed from α. Thus, there is an algorithm to compute the bigraded minimal free resolution of a set of fat points Z which satisfies Convention 7.1. For the convenience of the reader, we explicitly write out this algorithm.

Algorithm 7.24 (Computing bigraded resolution).

Input: $\alpha = (\alpha_1,\ldots,\alpha_h)$ with $\alpha_1 \geq \alpha_2 \geq \cdots \geq \alpha_h$ where α describes the ACM support of Z.
Output: The shifts in the bigraded minimal free resolution of $I(Z)$.

Step 1: Compute the shifts in the bigraded resolution of $I(Y)$ where Y is the completion of Z.

- $SY_0 := \{(2h,0),(h,\alpha_1),(0,2\alpha_1)\} \cup \{(i-1,\alpha_1+\alpha_i),(i+h-1,\alpha_i) \mid \alpha_i - \alpha_{i-1} < 0\}$
- $SY_1 := \{(2h,\alpha_h),(h,\alpha_1+\alpha_r)\} \cup \{(i-1,\alpha_1+\alpha_{i-1}),(i+h-1,\alpha_{i-1}) \mid \alpha_i - \alpha_{i-1} < 0\}$

Step 2: Locate the corners

- $C_0 := \{(i,\alpha_i+1) \mid \alpha_i - \alpha_{i-1} < 0\} = \{(i_1,j_1),\ldots,(i_s,j_s)\}$

(lexicographically ordered

from largest to smallest)

- $C_1 := \{(i_a,j_b) \mid (i_a,j_a),(i_b,j_b) \in C_0 \text{ and } a > b\}$
- $C := C_0 \cup C_1$ and order C in lexicographical order (largest to smallest)

Step 3: Calculate the shifts in the resolution of $I(Z)$.

- Let M_α be the $h \times \alpha_1$ matrix where $(M_\alpha)_{i,j} = \begin{cases} 2 \text{ if } j \leq \alpha_i \\ 1 \text{ otherwise} \end{cases}$
- Set $SZ_0 := SY_0$, $SZ_1 := SY_1$, and $SZ_2 := \{\}$
- For each $(i,j) \in C$ (working largest to smallest) do

$$u_{i,j} := (M_\alpha)_{1,j} + \cdots + (M_\alpha)_{i-1,j}$$
$$v_{i,j} := (M_\alpha)_{i,1} + \cdots + (M_\alpha)_{i,j-1}$$
$$a_{i,j} := (M_\alpha)_{i,j} + \cdots + (M_\alpha)_{h,j}$$
$$b_{i,j} := (M_\alpha)_{i,j} + \cdots + (M_\alpha)_{i,\alpha_1}$$
$$SZ_0 := SZ_0 \cup \{(u_{i,j}, v_{i,j})\}$$
$$SZ_1 := SZ_1 \cup \{(u_{i,j} + a_{i,j}, v_{i,j}), (u_{i,j}, v_{i,j} + b_{i,j})\}$$
$$SZ_2 := SZ_2 \cup \{(u_{i,j} + a_{i,j}, v_{i,j} + b_{i,j})\}$$
$$(M_\alpha)_{ij} := \begin{cases} 0 & \text{if } (i',j') \succeq (i,j) \\ (M_\alpha)_{ij} & \text{otherwise} \end{cases}$$

Step 4: Return SZ_0, SZ_1, and SZ_2 (the shifts at the 0-th, 1-st, and 2-nd step of the resolution, respectively).

Remark 7.25. The above algorithm has been implemented in CoCoA [1] and *Macaulay* 2 [41], and can be downloaded from the second author's web page.

Example 7.26. We use Algorithm 7.24 to compute the minimal bigraded free resolution of the ideal of fat points of Example 7.2. We have already computed SY_0 and SY_1 in Example 7.6. To calculate the remaining elements of SZ_0, SZ_1, and SZ_2, where SZ_i is the set of shifts in i-th free module appearing the resolution of $I(Z)$, we need the numbers $u_{i,j}, v_{i,j}, a_{i,j}, b_{i,j}$ for each corner $(i,j) \in C$. We have presented these numbers in the table below:

$(i,j) \in C$	$u_{i,j}$	$v_{i,j}$	$a_{i,j}$	$b_{i,j}$
(4,6)	4	6	2	1
(4,4)	5	4	2	2
(4,2)	6	2	2	2
(3,6)	3	8	1	1
(3,4)	4	6	1	2
(2,6)	2	10	1	1

By using Theorem 7.22 and the above information, we have

$$SZ_0 = \{(6,2),(5,4),(4,6),(4,6),(3,8),(2,10)\} \cup SY_0$$
$$SZ_1 = \{(8,2),(7,4),(6,6),(6,4),(5,6),(5,6),(4,8),(4,8),(4,7),(3,10),$$
$$(3,9),(2,11)\} \cup SY_1$$
$$SZ_2 = \{(8,4),(7,6),(6,7),(5,8),(4,9),(3,11)\}.$$

Remark 7.27. From Algorithm 7.24 we see that Z is ACM if and only if $C = \emptyset$ if and only if $\alpha = (\alpha_1,\ldots,\alpha_1)$, that is, if the support of Z is a complete intersection and $Z = Y$.

7.6 Additional notes

The notion of a completion and the forms F_ℓ relative to a corner (i,j) first appeared in the work of the first author [42, 43]. Specifically, the completion of a set of double points with ACM support was used to determine the minimal generators and bigraded Hilbert function of $I(Z)$. The main results of this chapter were first proved by the authors in [46]. Because we have computed the bigraded minimal free resolution of $I(Z)$, one can also adapt our algorithm to compute the bigraded Hilbert function of $I(Z)$.

The main result of this chapter is similar in spirit to the work of Catalisano [15] on fat points on a conic. Specifically, Catalisano produced an algorithmic procedure to find the minimal free resolution and the Hilbert function of the ideal of a set of fat points on a conic curve in $\mathbb{P}^2$.

In this chapter we have restricted to the case that the support of our double points is ACM. Some results are also known about the case that Z is a set of double fat points whose support is in generic position (recall that a set of points $X \subseteq \mathbb{P}^1 \times \mathbb{P}^1$ is in generic position if $H_X(i,j) = \min\{\dim_k R_{i,j}, |X|\}$ for all $(i,j) \in \mathbb{N}^2$). In particular, using work of Catalisano, Geramita, and Gimigliano [17], the second author determined the Hilbert function of Z when each point has multiplicity two and where the support of Z is in generic position (see [96]).

Using the correspondence between points in $\mathbb{P}^1 \times \mathbb{P}^1$ and special configurations of lines in $\mathbb{P}^3$ (see Remark 3.3), we can reinterpret the results of this chapter about "fat double lines" in $\mathbb{P}^3$. This point-of-view will be exploited in the next chapter. As a general comment, there appears to be little in the literature about the minimal free resolutions of unions of fat lines.

Chapter 8
Applications

In the previous chapter we described how to compute the bigraded minimal free resolution of $I(Z)$ when Z is a set of double points in $\mathbb{P}^1 \times \mathbb{P}^1$ with the property that $X = \mathrm{Supp}(Z)$ is ACM. In this situation the bigraded minimal free resolution of $I(Z)$ is a function of the tuple $\alpha_X = (\alpha_1, \ldots, \alpha_h)$ associated with X.

In this last chapter we exploit this bigraded minimal free resolution. As a first application we show how this resolution gives additional evidence to a question raised by Römer [84]. Römer's question asks how the i-th Betti of an ideal I is related to some partial information about the (k,j)-th graded Betti numbers $\beta_{k,j}(I)$ with $k \neq i$. As a second application, we use the results of the previous chapter to give a negative answer to a question of Huneke on symbolic powers. In both applications we utilize the fact that we can "coarsen" the bigrading of the ideal $I(Z)$ to deduce some results about the ideal $I(Z)$ when viewed as a graded ideal.

8.1 An application: Betti numbers and a question of Römer

Let I be a homogeneous ideal of $R = k[x_1, \ldots, x_n]$ and consider the graded minimal free resolution of R/I

$$0 \to \mathbb{F}_p \to \mathbb{F}_{p-1} \to \cdots \to \mathbb{F}_1 \to R \to R/I \to 0$$

where $\mathbb{F}_i = \bigoplus_{j \in \mathbb{Z}} R(-j)^{\beta_{i,j}(R/I)}$. The number $p = \text{proj-dim}(R/I)$ is the *projective dimension*, while the numbers $\beta_{i,j}(R/I)$ are the (i,j)-th *graded Betti numbers* of R/I.

Römer [84] initiated an investigation into the relationship between the *i-th Betti number of* R/I, i.e., $\beta_i(R/I) = \sum_{j \in \mathbb{Z}} \beta_{i,j}(R/I)$, and the shifts that appear with the minimal free resolution. Among other things, Römer asked the following question.

© The Authors 2015
E. Guardo, A. Van Tuyl, *Arithmetically Cohen-Macaulay Sets of Points in* $\mathbb{P}^1 \times \mathbb{P}^1$,
SpringerBriefs in Mathematics, DOI 10.1007/978-3-319-24166-1_8

Question 8.1. Let I be homogeneous ideal of $R = k[x_0,\dots,x_n]$. Does the following bound hold for all $i = 1,\dots,p$:

$$\beta_i(R/I) \leq \frac{1}{(i-1)!(p-i)!}\prod_{j\neq i} M_j$$

where $M_i := \max\{j \mid \beta_{i,j}(R/I) \neq 0\}$ denotes the maximum shift that appears in $\mathbb{F}_i$.

In this section, we show that the ideals $I(Z)$ studied in the previous chapter satisfy Question 8.1:

Theorem 8.2. *Let Z be a set of double points in $\mathbb{P}^1 \times \mathbb{P}^1$ with ACM support. Then all the i-th Betti numbers of $R/I(Z)$ satisfy the bound of Question 8.1.*

Although we have viewed $R/I(Z)$ as a bigraded ring up to this point, the ring $R/I(Z)$ also can be given a graded structure by defining the i-th graded piece to be $(R/I(Z))_i = \bigoplus_{a+b=i}(R/I(Z))_{a,b}$ (see also Remark 3.3). As pointed out in Remark 7.27, $R/I(Z)$ is rarely Cohen-Macaulay, so this family provides further evidence that Question 8.1 holds for all codimension 2 ideals (Römer showed Question 8.1 is true for all codimension 2 Cohen-Macaulay ideals).

Going forward, we continue to use the notation developed in Chapter 7. In particular, we continue to assume Z satisfies Convention 7.1. We first show how to obtain precise formulas for $\beta_i(R/I(Z))$ for $i = 1,2$ and 3, and lower bounds for M_1, M_2 and M_3 using α. With this information, the verification of the bound in Question 8.1 is a straightforward exercise.

Let $\alpha = (\alpha_1,\dots,\alpha_h)$ be any partition, i.e., $\alpha_1 \geq \alpha_2 \geq \cdots \geq \alpha_h \geq 1$. We set

$$d(\alpha) := \#\{i \mid \alpha_i - \alpha_{i-1} < 0\}.$$

Also, let $i^\star := \min\{i \mid \alpha_i - \alpha_{i-1} < 0\}$. This means $\alpha_1 = \alpha_2 = \cdots = \alpha_{i^\star-1} > \alpha_{i^\star}$.

Lemma 8.3. *Let Z be a set of double points in $\mathbb{P}^1 \times \mathbb{P}^1$ such that $X = \mathrm{Supp}(Z)$ is ACM. Let $\alpha_X = (\alpha_1,\dots,\alpha_h)$ be the tuple associated with X. If $d = d(\alpha_X)$, then*

(i) $\beta_1(R/I(Z)) = 2d+3+\binom{d+1}{2}$.
(ii) $\beta_2(R/I(Z)) = 2d+2+2\binom{d+1}{2}$.
(iii) $\beta_3(R/I(Z)) = \binom{d+1}{2}$.

Proof. Let Y be the completion of Z. By Theorem 7.4, $R/I(Y)$ is ACM, and $\beta_1(R/I(Y)) = 3+2d$ and $\beta_2(R/I(Y)) = 2+2d$. By Theorem 7.12 there exist p forms $F_1,\dots,F_p$ such that $I(Z) = I(Y)+(F_1,\dots,F_p)$. Here, p is the number of corners which is $p = \binom{d+1}{2}$. So $\beta_1(R/I(Z)) = 2d+3+\binom{d+1}{2}$. By Theorem 7.22, each generator F_i contributes two first syzygies and one second syzygy. Hence $\beta_2(R/I(Z)) = 2d+2+2\binom{d+1}{2}$ and $\beta_3(R/I(Z)) = \binom{d+1}{2}$.

Lemma 8.4. *Let Z be a set of double points in $\mathbb{P}^1 \times \mathbb{P}^1$ such that $X = \mathrm{Supp}(Z)$ is ACM. Let $\alpha_X = (\alpha_1, \ldots, \alpha_h)$ be the tuple associated with X. Suppose that $d(\alpha_X) > 0$. Then*

(i) $2\alpha_1 \leq M_1$.
(ii) $2\alpha_1 + 1 \leq M_2$.
(iii) $\alpha_1 + \alpha_{i^\star} + 3 \leq M_3$.

Proof. Let Y be the completion of Z. By Theorem 7.4 there is a generator of $I(Y)$ of bidegree $(0, 2\alpha_1)$ and a first syzygy of $I(Y)$ of bidegree $(i^\star - 1, \alpha_1 + \alpha_{i^\star-1})$. By Algorithm 7.24 we thus have that the bigraded shift $(0, -2\alpha_1)$ appears as a shift in $\mathbb{F}_1$ and $(-i^\star + 1, -\alpha_1 - \alpha_{i^\star-1})$ appears as a shift in $\mathbb{F}_2$. So, if we only consider the graded resolution of $R/I(Z)$, we have that there must be a shift of $-2\alpha_1$ in $\mathbb{F}_1$ and a shift of $-i^\star + 1 - \alpha_1 - \alpha_{i^\star-1} \leq -1 - \alpha_1 - \alpha_1$ in $\mathbb{F}_2$. So $M_1 \geq 2\alpha_1$ and $M_2 \geq 2\alpha_1 + 1$.

Note that $(i^\star, \alpha_{i^\star} + 1)$ is a base corner of Z, and in fact, is the smallest corner of Z with respect to the lexicographical ordering. Consider the matrix Mx_p as defined before Lemma 7.21. It must have the following form:

$$\begin{bmatrix} 2\,2 \cdots 2\,2 \cdots 2 \\ \vdots \\ 2\,2 \cdots 2\,1 \cdots 1 \\ \vdots \end{bmatrix}.$$

That is, the first row contains α_1 twos, and row $i^\star$ contains $\alpha_i^\star$ twos and $\alpha_1 - \alpha_{i^\star}$ ones. By Theorem 7.22 there is a second syzygy of $I(Z)$ whose bidegree is (u, v) where u is the sum of the entries in column $\alpha_{i^\star} + 1$ and v is the sum of the entries in row $i^\star$ of the above matrix. Hence $u \geq 2 + 1$ and $v = 2\alpha_i^\star + (\alpha_1 - \alpha_{i^\star}) = \alpha_1 + \alpha_{i^\star}$. So, in the graded resolution of $R/I(Z)$, there is a shift of $-u - v \leq -3 - \alpha_1 - \alpha_i^\star$, from which we deduce that $M_3 \geq \alpha_1 + \alpha_i^\star + 3$.

With the above lemmas, we now prove Theorem 8.2.

Proof (of Theorem 8.2). Let $\alpha_X = (\alpha_1, \ldots, \alpha_h)$ be the tuple associated with the support of Z, and set $d = d(\alpha_X)$. If $d = 0$, then $\alpha_X = (\alpha_1, \ldots, \alpha_1)$, and in this case $R/I(Z)$ is Cohen-Macaulay of codimension 2, and thus satisfies the bound in Question 8.1 by [84, Corollary 4.2].

So, we can assume that $d \geq 1$. Now $R/I(Z)$ is not ACM because $\beta_3(R/I(Z)) = \binom{d+1}{2} > 0$. Before proceeding, we note that $\alpha_1 - 1 \geq d$ and $\alpha_{i^\star} \geq d$. We need to verify the bound of Question 8.1 for $i = 1, 2$ and 3 where $p = 3$ in this case. We consider each case separately.

Case: $i = 1$. In this case, we have

$$\beta_1(R/I(Z)) = 2d + 3 + \binom{d+1}{2} = \frac{1}{2}(d+2)(d+3).$$

But $(d+2) \leq (2d+3)$ and $(d+3) \leq (2d+3)$ for all $d \geq 1$, so

$$\begin{aligned}\beta_1(R/I(Z)) &\leq \frac{1}{2}(2d+3)(2d+3) \leq \frac{1}{2}(2(d+1)+1)((d+1)+d+2)\\ &\leq \frac{1}{2}(2\alpha_1+1)(\alpha_1+\alpha_{i^\star}+2) \leq \frac{1}{(1-1)!(3-1)!}M_2M_3.\end{aligned}$$

Case: $i = 2$. For this case

$$\begin{aligned}\beta_2(R/I(Z)) &= 2d+2+2\binom{d+1}{2} = 2d+2+(d+1)d = (d+1)(d+2)\\ &\leq 2(d+1)(d+3)\\ &\leq (2(d+1))(2(d+2)) = (2(d+1))((d+1)+d+3)\\ &\leq (2\alpha_1)(\alpha_1+\alpha_{i^\star}+3) \leq \frac{1}{(2-1)!(3-2)!}M_1M_3.\end{aligned}$$

Case: $i = 3$. In our final case we have

$$\begin{aligned}\beta_3(R/I(Z)) &= \binom{d+1}{2} \leq \binom{\alpha_1+1}{2} \leq \alpha_1(\alpha_1+1)\\ &\leq \alpha_1(2\alpha_1+1) = \frac{1}{2}2\alpha_1(2\alpha_1+1) \leq \frac{1}{(3-1)!(3-3)!}M_1M_2.\end{aligned}$$

So, the bound of Question 8.1 is satisfied for all $i = 1,2,3$.

8.2 An application: symbolic powers and a question of Huneke

As in the previous section, we apply the results of Chapter 7 to a current research programme. Specifically, we will present a result on symbolic powers of $I(X)$ when X is an ACM set of points in $\mathbb{P}^1 \times \mathbb{P}^1$. As a consequence, we will be able to give a negative answer to a question of C. Huneke (see Remark 8.9 and Example 8.14).

We start with a brief digression on symbolic powers of ideals. Let P be any prime ideal in R, and let

$$\varphi : R \to R_P$$

be the natural ring homomorphism from R to R_P, that is, the ring R localized at the prime ideal P. If we set $Q = P^m$, then QR_P denotes the ideal generated by Q in R_P.

Definition 8.5. Let P be a prime ideal of R. With the notation as above, the *m-th symbolic power of* P, denoted $P^{(m)}$, is the ideal $\varphi^{-1}(QR_P)$ in R.

The above definition can be extended to any ideal by using the set of primes associated with the ideal. More precisely:

Definition 8.6. Suppose that I is an ideal of R whose set of associated primes is $\mathrm{Ass}(I) = \{P_1,\ldots,P_s\}$. Then then *m-th symbolic power of I*, denoted $I^{(m)}$, is the ideal of R given by

$$I^{(m)} := \bigcap_{P \in \mathrm{Ass}(I)} \varphi_P^{-1}(I^m R_P)$$

where φ_P is the natural ring homomorphism $\varphi_P : R \to R_P$.

In general, $I^m \subseteq I^{(m)}$. Although there are a number of natural problems about this containment, we highlight the following two questions:

Question 8.7. Let I be an ideal of R.

(*i*) When is $I^m = I^{(m)}$ for all $m \geq 1$?
(*ii*) Is there an M such that $I^m = I^{(m)}$ for all $m \geq 1$ if $I^m = I^{(m)}$ for $1 \leq m \leq M$?

We will answer both of these question for the case that $I = I(X)$ when X is an ACM set of points in $\mathbb{P}^1 \times \mathbb{P}^1$. In particular, we will prove the following theorem.

Theorem 8.8. *Suppose that X is an ACM set of points in $\mathbb{P}^1 \times \mathbb{P}^1$. If $I = I(X)$, then $I^{(m)} = I^m$ for all $m \geq 1$ if and only if $I^{(3)} = I^3$.*

Remark 8.9. Craig Huneke, in a talk entitled “Comparing Powers and Symbolic Powers of Ideals” given at the University of Nebraska, Lincoln in May 2008, raised the question of whether the value of M in Question 8.7 (*ii*) can be taken to be the *bigheight* of I, that is, the maximal height among the heights of the associated primes of I. Theorem 8.8 gives a negative answer to this question. To see why, note that all the associated primes of $I(X)$ have height two, so the bigheight of $I(X)$ is two. However, in order to verify that $I(X)^m = I(X)^{(m)}$ for all $m \geq 1$, we need to verify the case $m = 3$, which is larger than the bigheight of $I(X)$.

We make the following observation which follows directly from the definitions.

Lemma 8.10. *Suppose that $X = \{P_1,\ldots,P_s\}$ is a set of points in $\mathbb{P}^1 \times \mathbb{P}^1$. Then*

$$I(X)^{(m)} = I(P_1)^m \cap I(P_2)^m \cap \cdots \cap I(P_s)^m \quad \textit{for all } m \geq 1.$$

To prove Theorem 8.8, we first prove the following interesting result that $I(X)^2 = I(X)^{(2)}$ for all ACM sets of points in $\mathbb{P}^1 \times \mathbb{P}^1$. Recall that when X is an ACM set of points in $\mathbb{P}^1 \times \mathbb{P}^1$, we can write down the generators of $I(X)$ using Corollary 5.6. Consequently, we can write down the generators of $I(X)^2$ for any finite set of ACM points X in $\mathbb{P}^1 \times \mathbb{P}^1$. On the other hand, by Lemma 8.10 the ideal $I(X)^{(2)}$ defines a set of fat points whose support X is ACM and whose points all have multiplicity two (alternatively, when viewed as a graded ideal, $I(X)^{(2)}$ defines a union of “fat lines” in $\mathbb{P}^3$). As a consequence, we can apply Algorithm 7.24 to $I(X)^{(2)}$. For our purposes, it is sufficient to know that the algorithm described in Algorithm 7.24 always produces a set of generators for $I(X)^{(2)}$ of the following form.

Lemma 8.11. *Suppose that X is an ACM set of points in $\mathbb{P}^1 \times \mathbb{P}^1$. Let $H_1,\dots,H_h$ denote the horizontal rules and $V_1,\dots,V_v$ denote the vertical rules which minimally contain X. There is a minimal set of generators of $I(X)^{(2)}$ such that every generator F has one of the following forms:*

1) $H_1^2 \cdots H_h^2$;
2) $H_1 \cdots H_h V_1 \cdots V_v$;
3) $V_1^2 \cdots V_v^2$;
4) or there exist $1 \le a \le b \le h$ and $1 \le c \le d \le v$ such that

$$F = H_1^2 H_2^2 \cdots H_a^2 H_{a+1}^1 \cdots H_b^1 V_1^2 V_2^2 \cdots V_c^2 V_{c+1}^1 \cdots V_d^1.$$

(If $a = b$ in case 4), F has the form $H_1^2 H_2^2 \cdots H_a^2 V_1^2 V_2^2 \cdots V_c^2 V_{c+1}^1 \cdots V_d^1$, and similarly for $c = d$.)

Proof (Sketch of the proof). For a minimal set of generators for $I(X)^{(2)}$ in terms of the partition α, see [43, Theorem 3.15] and Algorithm 7.24; in particular, explicit values of a,b,c, and d are given for the elements described in 4.

Theorem 8.12. *Suppose that X is an ACM set of points in $\mathbb{P}^1 \times \mathbb{P}^1$. Then $I(X)^2 = I(X)^{(2)}$.*

Proof. Let $I = I(X)$. It suffices to prove $I^{(2)} \subseteq I^2$. Let F be any generator of $I^{(2)}$. By Lemma 8.11, F must have one of four forms. Since $H_1 \cdots H_h$ and $V_1 \cdots V_v$ are generators of I, the generators $H_1^2 \cdots H_h^2$, $H_1 \cdots H_h V_1 \cdots V_v$, and $V_1^2 \cdots V_v^2$ all belong to I^2.

We therefore take a generator of $I^{(2)}$ of the form

$$F = H_1^2 H_2^2 \cdots H_a^2 H_{a+1}^1 \cdots H_b^1 V_1^2 V_2^2 \cdots V_c^2 V_{c+1}^1 \cdots V_d^1$$

for some $1 \le a \le b \le h$ and $1 \le c \le d \le v$. Factor F as

$$F_1 = H_1 H_2 \cdots H_a V_1 \cdots V_c V_{c+1} \cdots V_d \text{ and } F_2 = H_1 H_2 \cdots H_a H_{a+1} \cdots H_b V_1 \cdots V_c.$$

We claim that both F_1 and F_2 are elements of I (and hence $F = F_1 F_2 \in I^2$). We show that $F_1 \in I$, since the other case is similar. Let $A \times B \in X$. Then $A \times B$ is either on one of the rulings $H_1,\dots,H_a$ or it is not. If it is on one of these rulings, say H_i, then $F_1(A \times B) = 0$ because $H_i(A \times B) = 0$. On the other hand, suppose that $A \times B$ is not on any of these rulings. Because F vanishes with multiplicity two at $A \times B$, and because $A \times B$ can lie on at most one of the rulings $H_{a+1},\dots,H_b$, there must be at least one vertical ruling V_j among $V_1,\dots,V_d$ such that $V_j(A \times B) = 0$. But this means $F_1(A \times B) = 0$. Hence $F_1 \in I$.

Example 8.13. The following example will help crystallize some of the above ideas. The ACM set of points $X \subseteq \mathbb{P}^1 \times \mathbb{P}^1$ of Example 3.16 has $\alpha_X = (6,5,3,1,1)$. By Corollary 5.6, the ideal $I(X)$ has minimal generators

$$\{V_{B_1}V_{B_2}V_{B_3}V_{B_4}V_{B_5}V_{B_6},\ H_{A_1}V_{B_1}V_{B_2}V_{B_3}V_{B_4}V_{B_5},\ H_{A_1}H_{A_2}V_{B_1}V_{B_2}V_{B_3},\ H_{A_1}H_{A_2}H_{A_3}V_{B_1},\ H_{A_1}H_{A_2}H_{A_3}H_{A_4}H_{A_5}\}.$$

The ideal $I(X)^2$ is then generated by all forms G_1G_2 where G_1 and G_2 are two not necessarily distinct elements from the above list. So, the ideal $I(X)^2$ will be generated by forms of degree

$$\{(10,0),(8,1),(7,3),(6,5),(6,2),(5,6),(5,4),(4,6),(4,6),(3,8),$$
$$(3,7),(2,10),(2,9),(1,11),(0,12)\}.$$

But this is the exact same degrees as the generators of $I(X)^{(2)}$ in Example 7.26, so $I(X)^2 = I(X)^{(2)}$.

We now present the proof of the main result of this section.

Proof (Proof of Theorem 8.8). Because I is homogeneous, we have $I^{(m)} = I^m$ if and only if $J^{(m)} = J^m$, where $J = IR_{\mathbf{m}}$, $R_{\mathbf{m}}$ being the localization of $R = k[\mathbb{P}^1 \times \mathbb{P}^1]$ at the ideal $\mathbf{m}$ generated by the variables. Note that J is a perfect ideal (i.e., $\text{proj-dim}_{R_M}(R_{\mathbf{m}}/J) = \text{depth}(J, R_M)$; we have $\text{depth}(J, R_{\mathbf{m}}) = \text{codim}(J) = 2$ since $R_{\mathbf{m}}$ is Cohen-Macaulay, and we obtain $\text{proj-dim}_{R_{\mathbf{m}}}(R_{\mathbf{m}}/J) = 2$ from the Auslander-Buchsbaum formula), it has codimension 2, $R_{\mathbf{m}}/J$ is Cohen-Macaulay, and J is generically a complete intersection (i.e., the localizations of J at its minimal associated primes are complete intersections). Now [76, Theorem 3.2] asserts that $J^{(m)} = J^m$ for all $m \geq 1$ if and only if $J^{(m)} = J^m$ for $1 \leq m \leq \dim(R_{\mathbf{m}}) - 1$. Because $\dim(R_{\mathbf{m}}) = 4$, it follows that $J^{(2)} = J^2$ and $J^{(3)} = J^3$ implies $J^{(m)} = J^m$ for all $m \geq 1$, and thus $I^{(2)} = I^2$ and $I^{(3)} = I^3$ implies $I^{(m)} = I^m$ for all $m \geq 1$. But we always have $I^{(2)} = I^2$ by Theorem 8.12, so the conclusion follows.

As the next two examples show, it is possible to find examples of ACM sets of points X in $\mathbb{P}^1 \times \mathbb{P}^1$ where $I(X)^{(3)} \neq I(X)^3$ and examples where $I(X)^{(3)} = I(X)^3$.

Example 8.14. Let X be an ACM set of points in $\mathbb{P}^1 \times \mathbb{P}^1$ with $\alpha_X = (3,2,1)$. This set of points can be visualized as in Figure 8.1. Let $I = I(X)$ be the ideal of X and let $\alpha(I) = \min\{d \mid I_d \neq (0)\}$, where I_d is the homogeneous component of I of degree d (with respect to the usual grading on $R = k[\mathbb{P}^3]$). Using Theorem 5.3, it is easy to check that $\alpha(I) = 3$ and hence $\alpha(I^3) = 3\alpha(I) = 9$. Thus, if we consider I^3 as a bigraded ideal, the bigraded component $(I^3)_{(4,4)}$ of bidegree $(4,4)$ is (0) (since $(I^3)_{(4,4)} \subseteq (I^3)_8 = (0)$). But there is a curve $C \subset \mathbb{P}^1 \times \mathbb{P}^1$ of bidegree $(4,4)$ vanishing

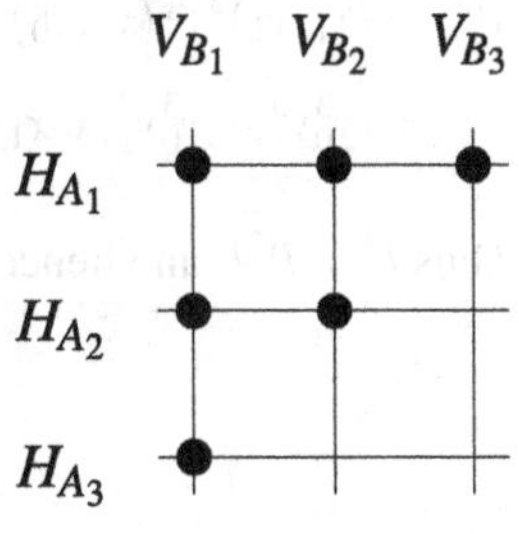

Fig. 8.1 An ACM set of points with $\alpha_X = (3,2,1)$

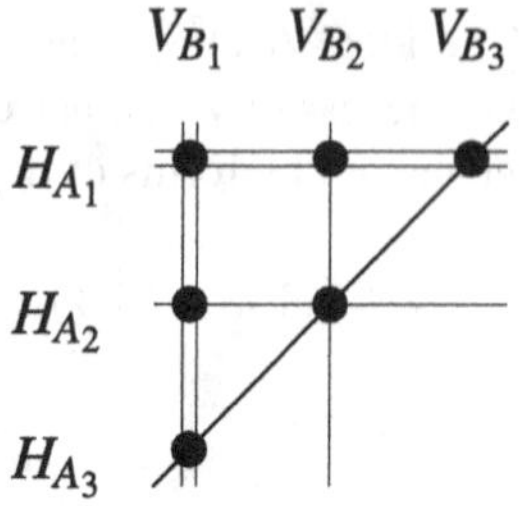

Fig. 8.2 The curve $H_{A_1}^2 V_{B_1}^2 H_{A_2} V_{B_2} F$ through X.

on X to order 3. It is, as shown in Figure 8.2, the zero locus of $H_{A_1}^2 V_{B_1}^2 H_{A_2} V_{B_2} F$, where $\deg F = (1,1)$, with the zero locus of F represented by the diagonal line. (One can check by Bézout's Theorem that in fact C is unique.) So, $(I^{(3)})_{(4,4)} \neq (0)$, whence $I^{(3)} \neq I^3$. This example gives a negative answer to the question of Huneke discussed in Remark 8.9.

A classical result of Zariski and Samuel [100, Lemma 5, Appendix 6] states that when I is a complete intersection, we always have $I^{(m)} = I^m$ for all positive integers m. So, when X is a complete intersection in $\mathbb{P}^1 \times \mathbb{P}^1$, we will always have $I(X)^{(m)} = I(X)^m$ for all $m \geq 1$. As shown in the next example, Theorem 8.8, enables us to show that there exist ACM sets of points in $\mathbb{P}^1 \times \mathbb{P}^1$ that are not complete intersections, but whose ideal $I(X)$ nevertheless has $I(X)^{(3)} = I(X)^3$, (and consequently, we have $I(X)^{(m)} = I(X)^m$ for all $m \geq 1$).

Example 8.15. Consider the ACM set of points

$$X = \{A_1 \times B_1, A_1 \times B_2, A_2 \times B_1\} \subseteq \mathbb{P}^1 \times \mathbb{P}^1$$

where $A_1 = B_1 = [1:0]$ and $A_2 = B_2 = [0:1] \in \mathbb{P}^1$. Note that $\alpha_X = (2,1)$, so by Theorem 5.13 the set X is not a complete intersection.

The ideal $I = I(X)$ of X is the monomial ideal $I = (x_1, y_1) \cap (x_1, y_0) \cap (x_0, y_1) = (x_0x_1, y_0y_1, x_0y_0)$. Then

$$I^3 = (y_0^3y_1^3, x_0y_0^3y_1^2, x_0x_1y_0^2y_1^2, x_0^2y_0^3y_1, x_0^2x_1y_0^2y_1, x_0^2x_1^2y_0y_1, x_0^3y_0^3, x_0^3x_1y_0^2, x_0^3x_1^2y_0, x_0^3x_1^3).$$

A tedious but elementary computation shows that

$$\begin{aligned} I^{(3)} &= (x_1, y_1)^3 \cap (x_1, y_0)^3 \cap (x_0, y_1)^3 \\ &= (y_0^3y_1^3, x_0y_0^3y_1^2, x_0x_1y_0^2y_1^2, x_0^2y_0^3y_1, x_0^2x_1y_0^2y_1, x_0^2x_1^2y_0y_1, x_0^3y_0^3, x_0^3x_1y_0^2, x_0^3x_1^2y_0, x_0^3x_1^3). \end{aligned}$$

Thus $I^3 = I^{(3)}$, and hence $I^m = I^{(m)}$ for all $m \geq 1$ by Theorem 8.8.

8.3 Additional notes

The results of Section 8.1 first appeared in a paper of the authors [46]. The original question of Römer appeared in [84]. Römer's original question was inspired by the Multiplicity Conjecture of Herzog, Huneke, and Srinivasan (see [63]). The Multiplicity Conjecture was eventually proved within the framework of what is now called Boij-Söderberg theory. We point the reader to the survey of Fløystad [26] for an introduction to this exciting area.

Symbolic powers of ideals is a topic at the forefront of current research in commutative algebra. Besides the questions listed in Question 8.7, the question of ideal containment has been of great interest. The ideal containment problem asks what is the smallest integer m such that $I^{(m)} \subseteq I^r$ for a fixed positive integer r. The papers of Bocci-Harbourne [5, 6], Ein-Lazarsfeld-Smith [23], Harbourne-Huneke [59], and Hochster-Huneke [65] form a very non-exhaustive list of papers on this interesting topic.

The material of Section 8.2 first appeared in a joint paper of the authors with B. Harbourne [53]. As we showed in Theorem 8.8, in order to determine if $I(X)^{(m)} = I(X)^m$ for all m when X is ACM, it suffices to check equality when $m = 3$. Because X is ACM, Theorem 4.11 implies X is described by α_X. It is therefore natural to ask if we can determine when $I(X)^{(3)} = I(X)^3$ in terms of α_X. This question was partially answered in [53]. In particular, evidence was presented for the following conjecture:

Conjecture 8.16. *Let X be a finite set of ACM points in $\mathbb{P}^1 \times \mathbb{P}^1$ with $\alpha_X = (\alpha_1, \ldots, \alpha_h)$. Let $I = I(X)$. Then $I^{(3)} = I^3$ if and only if α_X has one of the following two forms:*

(i) $\alpha_X = (a, a, \ldots, a)$ *for some integer* $a \geq 1$.
(ii) $\alpha_X = (a, \ldots, a, b, \ldots, b)$ *for some integers* $a > b \geq 1$.

When $\alpha_X = (a, \ldots, a)$, then Theorem 5.13 implies that X is a complete intersection, so it is true that $I(X)^{(3)} = I(X)^3$. In early 2015, as we were writing the final version of this monograph, this conjecture was investigated at the mini-workshop "Ideals of Linear Subspaces, Their Symbolic Powers and Waring Problems" which was held at Mathematisches Forschungsinstitut Oberwolfach (see the Oberwolfach Report [8]). Juan Migliore observed that the work of Chris Peterson [81] can be used to finish one direction of Conjecture 8.16. In particular, it was observed that when $\alpha_X = (a, \ldots, a, b, \ldots, b)$, then $I(X)$ is a quasi-complete intersection (it is generated by three elements). So, the fact $I(X)^3 = I(X)^{(3)}$ then follows from [81, Corollary 2.7] because $I(X)$ also defines a curve in $\mathbb{P}^3$ that is a local complete intersection. A possible strategy to complete the proof of Conjecture 8.16 was also suggested and is currently being explored by the authors and other participants of the workshop.

In keeping with the theme of this monograph, we have only considered symbolic powers of points of X in $\mathbb{P}^1 \times \mathbb{P}^1$ when X is ACM. However, some other work has been carried out in the case that the support is in generic position. In particular, see the paper of Baczyńska-Dumnicki-Habura-Malara-Pokora-Szemberg-Szpond-Tutaj-Gasińska [3] and a paper of the authors and Harbourne [54].

References

1. J. Abbott, A.M. Bigatti, G. Lagorio, CoCoA-5: a system for doing computations in commutative algebra. Available at http://cocoa.dima.unige.it
2. A. Aramova, K. Crona, E. De Negri, Bigeneric initial ideals, diagonal subalgebras and bigraded Hilbert functions. J. Pure Appl. Algebra **150**(3), 215–235 (2000)
3. M. Baczyńska, M. Dumnicki, A. Habura, G. Malara, P. Pokora, T. Szemberg, J. Szpond, H. Tutaj-Gasińska, Points fattening on $\mathbb{P}^1 \times \mathbb{P}^1$ and symbolic powers of bi-homogeneous ideals. J. Pure Appl. Algebra **218**(8), 1555–1562 (2014)
4. E. Ballico, Postulation of disjoint unions of lines and a few planes. J. Pure Appl. Algebra **215**(4), 597–608 (2011)
5. C. Bocci, B. Harbourne, Comparing powers and symbolic power of ideals. J. Algebraic Geom. **19**(3), 399–417 (2010)
6. C. Bocci, B. Harbourne, The resurgence of ideals of points and the containment problem. Proc. Am. Math. Soc. **138**(4), 1175–1190 (2010)
7. C. Bocci, R. Miranda, Topics on interpolation problems in algebraic geometry. Rend. Sem. Mat. Univ. Politec. Torino **62**(4), 279–334 (2004)
8. C. Bocci, E. Carlini, B. Harbourne, E. Guardo, Mini-workshop: ideals of linear subspaces, their symbolic powers and waring problems. Oberwolfach Rep., vol. 12, no. 1 (2015)
9. P. Bonacini, L. Marino, On the Hilbert function of zero-dimensional schemes in $\mathbb{P}^1 \times \mathbb{P}^1$. Collect. Math. **62**(1), 57–67 (2011)
10. P. Bonacini, L. Marino, Hilbert functions and set of points in $\mathbb{P}^1 \times \mathbb{P}^1$. Beitr. Algebra Geom. **56**(1), 43–61 (2015)
11. P. Bonacini, L. Marino, Minimal free resolutions of zero-dimensional schemes in $\mathbb{P}^1 \times \mathbb{P}^1$. Algebra Colloq. **22**(1), 97–108 (2015)
12. W. Bruns, J. Herzog, *Cohen-Macaulay Rings (Revised Version)* (Cambridge University Press, New York, 1998)
13. E. Carlini, M.V. Catalisano, A.V. Geramita, Bipolynomial Hilbert functions. J. Algebra **324**(4), 758–781 (2010)
14. E. Carlini, M.V. Catalisano, A.V. Geramita, Subspace arrangements, configurations of linear spaces and the quadrics containing them. J. Algebra **362**, 70–83 (2012)
15. M.V. Catalisano, "Fat" points on a conic. Commun. Algebra **19**(8), 2153–2168 (1991)
16. M.V. Catalisano, A.V. Geramita, A. Gimigliano, Ranks of tensors, secant varieties of Segre varieties and fat points. Linear Algebra Appl. **355**(1), 263–285 (2002)
17. M.V. Catalisano, A.V. Geramita, A. Gimigliano, Higher secant varieties of Segre-Veronese varieties, in *Projective Varieties with Unexpected Properties* (Walter de Gruyter, Berlin, 2005), pp. 81–107

© The Authors 2015

E. Guardo, A. Van Tuyl, *Arithmetically Cohen-Macaulay Sets of Points in* $\mathbb{P}^1 \times \mathbb{P}^1$,
SpringerBriefs in Mathematics, DOI 10.1007/978-3-319-24166-1

18. L. Chiantini, D. Sacchi, Segre functions in multiprojective spaces and tensor analysis. Trends Hist. Sci. (2014, to appear)
19. D. Cox, J. Little, D. O'Shea, *Ideals, Varieties, and Algorithms. An Introduction to Computational Algebraic Geometry and Commutative Algebra*. Undergraduate Texts in Mathematics (Springer, New York, 1992)
20. D. Cox, J. Little, D. O'Shea, *Using Algebraic Geometry*. Graduate Texts in Mathematics, vol. 185 (Springer, New York, 1998)
21. D. Cox, A. Dickenstein, H. Schenck, A case study in bigraded commutative algebra, in *Syzygies and Hilbert Functions*. Lecture Notes in Pure and Applied Mathematics, vol. 254 (Chapman & Hall/CRC, Boca Raton, 2007), pp. 67–111
22. K. Crona, Standard bigraded Hilbert functions. Commun. Algebra **34**(2), 425–462 (2006)
23. L. Ein, R. Lazarsfeld, K.E. Smith, Uniform bounds and symbolic powers on smooth varieties. Invent. Math. **144**(2), 241–252 (2001)
24. D. Eisenbud, *The Geometry of Syzygies. A Second Course in Commutative Algebra and Algebraic Geometry*. Graduate Texts in Mathematics, vol. 229 (Springer, New York, 2005)
25. D. Eisenbud, S. Popescu, Gale duality and free resolutions of ideals of points. Invent. Math. **136**(2), 419–449 (1999)
26. G. Fløystad, Boij-Söderberg theory: introduction and survey, in *Progress in Commutative Algebra 1* (de Gruyter, Berlin, 2012), pp. 1–54
27. C. Francisco, A. Van Tuyl, Some families of componentwise linear monomial ideals. Nagoya Math. J. **187**, 115–156 (2007)
28. A.V. Geramita, Zero-dimensional schemes: singular curves and rational surfaces, in *Zero-Dimensional Schemes (Ravello, 1992)* (de Gruyter, Berlin, 1994), pp. 1–10
29. A.V. Geramita, P. Maroscia, The ideal of forms vanishing at a finite set of points in $\mathbb{P}^n$. J. Algebra **90**(2), 528–555 (1984)
30. A.V. Geramita, H. Schenck, Fat points, inverse systems, and piecewise polynomial functions. J. Algebra **204**(1), 116–128 (1998)
31. A.V. Geramita, P. Maroscia, L.G. Roberts, The Hilbert function of a reduced k-algebra. J. Lond. Math. Soc. (2) **28**(3), 443–452 (1983)
32. A.V. Geramita, D. Gregory, L. Roberts, Monomial ideals and points in projective space. J. Pure Appl. Algebra **40**(1), 33–62 (1986)
33. A.V. Geramita, M. Kreuzer, L. Robbiano, Cayley-Bacharach schemes and their canonical modules. Trans. Am. Math. Soc. **339**(1), 163–189 (1993)
34. A.V. Geramita, J. Migliore, L. Sabourin, On the first infinitesimal neighborhood of a linear configuration of points in $\mathbb{P}^2$. J. Algebra **298**(2), 563–611 (2006)
35. A. Gimigliano, Our thin knowledge of fat points, in *The Curves Seminar at Queen's, Vol. VI (Kingston, ON, 1989)*. Queen's Papers in Pure and Applied Mathematics, vol. 83 (Queen's University, Kingston, 1989), Exp. No. B, 50 pp.
36. S. Giuffrida, R. Maggioni, A. Ragusa, On the postulation of 0-dimensional subschemes on a smooth quadric. Pac. J. Math. **155**(2), 251–282 (1992)
37. S. Giuffrida, R. Maggioni, A. Ragusa, Resolutions of zero-dimensional subschemes of a smooth quadric, in *Zero-Dimensional Schemes (Ravello, 1992)* (de Gruyter, Berlin, 1994), pp. 191–204
38. S. Giuffrida, R. Maggioni, A. Ragusa, Resolutions of generic points lying on a smooth quadric. Manuscripta Math. **91**(4), 421–444 (1996)
39. S. Giuffrida, R. Maggioni, G. Zappalà, Scheme-theoretic complete intersections in $\mathbb{P}^1 \times \mathbb{P}^1$. Commun. Algebra **41**(2), 532–551 (2013)
40. L. Gold, J. Little, H. Schenck, Cayley-Bacharach and evaluation codes on complete intersections. J. Pure Appl. Algebra **196**(1), 91–99 (2005)
41. D.R. Grayson, M.E. Stillman, Macaulay 2, a software system for research in algebraic geometry. http://www.math.uiuc.edu/Macaulay2/
42. E. Guardo, Schemi di "Fat Points". PhD Thesis, Università di Messina (2000)
43. E. Guardo, Fat point schemes on a smooth quadric. J. Pure Appl. Algebra **162**(2–3), 183–208 (2001)

44. E. Guardo, A survey on fat points on a smooth quadric, in *Algebraic Structures and Their Representations*. Contemporary Mathematics, vol. 376 (American Mathematical Society, Providence, 2005), pp. 61–87
45. E. Guardo, A. Van Tuyl, Fat Points in $\mathbb{P}^1 \times \mathbb{P}^1$ and their Hilbert functions. Can. J. Math. **56**(4), 716–741 (2004)
46. E. Guardo, A. Van Tuyl, The minimal resolutions of double points in $\mathbb{P}^1 \times \mathbb{P}^1$ with ACM support. J. Pure Appl. Algebra **211**(3), 784–800 (2007)
47. E. Guardo, A. Van Tuyl, Separators of points in a multiprojective space. Manuscripta Math. **126**(1), 99–113 (2008)
48. E. Guardo, A. Van Tuyl, ACM sets of points in multiprojective space. Collect. Math. **59**(2), 191–213 (2008)
49. E. Guardo, A. Van Tuyl, Separators of fat points in $\mathbb{P}^n \times \mathbb{P}^m$. J. Pure Appl. Algebra **215**(8), 1990–1998 (2011)
50. E. Guardo, A. Van Tuyl, Classifying ACM sets of points in $\mathbb{P}^1 \times \mathbb{P}^1$ via separators. Arch. Math. **99**(1), 33–36 (2012)
51. E. Guardo, A. Van Tuyl, Separators of arithmetically Cohen-Macaulay fatpoints in $\mathbb{P}^1 \times \mathbb{P}^1$. J. Commut. Algebra **4**(2), 255–268 (2012)
52. E. Guardo, A. Van Tuyl, On the Hilbert functions of sets of points in $\mathbb{P}^1 \times \mathbb{P}^1 \times \mathbb{P}^1$. Math. Proc. Camb. Philos. Soc. **159**(1), 115–123 (2015)
53. E. Guardo, B. Harbourne, A. Van Tuyl, Fat lines in $\mathbb{P}^3$: powers versus symbolic powers. J. Algebra **390**, 221–230 (2013)
54. E. Guardo, B. Harbourne, A. Van Tuyl, Symbolic powers versus regular powers of ideals of general points in $\mathbb{P}^1 \times \mathbb{P}^1$. Can. J. Math. **65**(4), 823–842 (2013)
55. H.T. Hà, Multigraded regularity, a^*-invariant and the minimal free resolution. J. Algebra **310**(1), 156–179 (2007)
56. H.T. Hà, A. Van Tuyl, The regularity of points in multi–projective spaces. J. Pure Appl. Algebra **187**(1–3), 153–167 (2004)
57. J. Hansen, Linkage and codes on complete intersections. Appl. Algebra Eng. Commun. Comput. **14**(3), 175–185 (2003)
58. B. Harbourne, Problems and progress: a survey on fat points in $\mathbb{P}^2$, in *Zero-Dimensional Schemes and Applications (Naples, 2000)* Queen's Papers in Pure and Applied Mathematics, vol. 123 (Queen's University, Kingston, 2002), pp. 85–132
59. B. Harbourne, C. Huneke, Are symbolic powers highly evolved? J. Ramanujan Math. Soc. **28A**, 247–266 (2013)
60. R. Hartshorne, *Algebraic Geometry*. Graduate Texts in Mathematics, vol. 52 (Springer, New York, 1977)
61. R. Hartshorne, A. Hirschowitz, Droites en position générale dans l'espace projectif, in *Algebraic Geometry (La Rábida, 1981)*. Lecture Notes in Mathematics, vol. 961 (Springer, Berlin, 1982), pp. 169–188
62. B. Hassett, *Introduction to Algebraic Geometry* (Cambridge University Press, Cambridge, 2007)
63. J. Herzog, H. Srinivasan, Bounds for multiplicities. Trans. Am. Math. Soc. **350**(7), 2879–2902 (1998)
64. A. Hirschowitz, C. Simpson, La résolution minimale de l'idéal d'un arrangement général d'un grand nombre de points dans $\mathbb{P}^n$. Invent. Math. **126**(3), 467–503 (1996)
65. M. Hochster, C. Huneke, Comparison of symbolic and ordinary powers of ideals. Invent. Math. **147**(2), 349–369 (2002)
66. J. Huizenga, Interpolation on surfaces in $\mathbb{P}^3$. Trans. Am. Math. Soc. **365**(2), 623–644 (2013)
67. A. Lorenzini, The minimal resolution conjecture. J. Algebra **156**(1), 5–35 (1993)
68. F.S. Macaulay, Some properties of enumeration in the theory of modular systems. Proc. Lond. Math. Soc. **26**(1), 531–555 (1927)
69. D. Maclagan, G.G. Smith, Multigraded Castelnuovo-Mumford regularity. J. Reine Angew. Math. **571**, 179–212 (2004)

70. L. Marino, Conductor and separating degrees for sets of points in $\mathbb{P}^r$ and in $\mathbb{P}^1 \times \mathbb{P}^1$. Boll. Unione Mat. Ital. Sez. B Artic. Ric. Mat. (8) **9**(2), 397–421 (2006)
71. L. Marino, A characterization of ACM 0-dimensional schemes in Q. Matematiche (Catania) **64**(2), 41–56 (2009)
72. J. Migliore, U. Nagel, Lifting monomial ideals. Special issue in honor of Robin Hartshorne. Comm. Algebra **28**(12), 5679–5701 (2000)
73. J. Migliore, U. Nagel, Liaison and related topics: notes from the Torino workshop-school. Rend. Sem. Mat. Univ. Politec. Torino **59**(2), 59–126 (2001)
74. E. Miller, B. Sturmfels, *Combinatorial Commutative Algebra*. Graduate Texts in Mathematics, vol. 227 (Springer, New York, 2005)
75. R. Miranda, Linear systems of plane curves. Not. Am. Math. Soc. **46**(2), 192–201 (1999)
76. S. Morey, Stability of associated primes and equality of ordinary and symbolic powers of ideals. Commun. Algebra **27**(7), 3221–3231 (1999)
77. F. Orecchia, Points in generic position and conductors of curves with ordinary singularities. J. Lond. Math. Soc. (2) **24**(1), 85–96 (1981)
78. G. Paxia, G. Raciti, A. Ragusa, Uniform position properties and Hilbert functions for points on a smooth quadric. J. Algebra **149**(1), 102–121 (1992)
79. I. Peeva, *Graded Syzygies*. Algebra and Applications, vol. 14 (Springer, London, 2011)
80. I. Peeva, M. Stillman, Open problems on syzygies and Hilbert functions. J. Commut. Algebra **1**(1), 159–195 (2009)
81. C. Peterson, Quasicomplete intersections, powers of ideals, and deficiency modules. J. Algebra **204**(1), 1–14 (1998)
82. G. Raciti, Hilbert function and geometric properties for a closed zero-dimensional subscheme of a quadric $Q \subseteq \mathbb{P}^3$. Commun. Algebra **18**(9), 3041–3053 (1990)
83. A. Ragusa, G. Zappalà, Postulation of subschemes of irreducible curves on a quadric surface. Rend. Circ. Mat. Palermo (2) **49**(1), 75–102 (2000)
84. T. Römer, Betti numbers and shifts in minimal graded free resolutions. Ill. J. Math. **54**(2), 449–467 (2010)
85. H.J. Ryser, Combinatorial Mathematics The Carus Mathematical Monographs, No. 14. Published by The Mathematical Association of America; distributed by Wiley, New York. (1963)
86. M. Şahin, I. Soprunov, Multigraded Hilbert function and toric complete intersection codes. Preprint (2014) [arXiv:1410.4164v2]
87. H. Schenck, *Computational Algebraic Geometry*. London Mathematical Society Student Texts, vol. 58 (Cambridge University Press, Cambridge, 2003)
88. R.Y. Sharp, *Steps in Commutative Algebra*. London Mathematical Society Student Texts, vol. 51, 2nd edn. (Cambridge University Press, Cambridge, 2000)
89. J. Sidman, A. Van Tuyl, Multigraded regularity: syzygies and fat points. Beitr. Algebra Geom. **47**(1), 67–87 (2006)
90. B.L. Van der Waerden, On Hilbert's function, series of composition of ideals and a generalisation of the theorem of Bezout. Proc. R. Acad. Amst. **31**, 749–770 (1928)
91. B.L. Van der Waerden, On varieties in multiple-projective spaces. Nederl. Akad. Wetensch. Indag. Math. **40**, 303–312 (1978)
92. A. Van Tuyl, Sets of Points in multi-projective spaces and their Hilbert function. PhD Thesis, Queen's University (2001).
93. A. Van Tuyl, The border of the Hilbert function of a set of points in $\mathbb{P}^{n_1} \times \cdots \times \mathbb{P}^{n_k}$. J. Pure Appl. Algebra **176**(2–3), 223–247 (2002)
94. A. Van Tuyl, The Hilbert functions of ACM sets of points in $\mathbb{P}^{n_1} \times \cdots \times \mathbb{P}^{n_k}$. J. Algebra **264**(2), 420–441 (2003)
95. A. Van Tuyl, The defining ideal of a set of points in multi-projective space. J. Lond. Math. Soc. (2) **72**(1), 73–90 (2005)

96. A. Van Tuyl, An appendix to a paper of M.V. Catalisano, A.V. Geramita, and A. Gimigliano: The Hilbert function of generic sets of 2-fat points in $\mathbb{P}^1 \times \mathbb{P}^1$, in *Projective Varieties with Unexpected Properties* (Walter de Gruyter, Berlin, 2005), pp. 109–112
97. O.L. Vidal, On the diagonals of a Rees algebra. PhD Thesis, Universitat de Barcelona (1999)
98. R.H. Villarreal, *Monomial Algebras*. Monographs and Textbooks in Pure and Applied Mathematics, vol. 238 (Marcel Dekker, New York, 2001)
99. G. Zappalà, 0-dimensional subschemes of curves lying on a smooth quadric surface. Matematiche (Catania) **52**(1), 115–127 (1997)
100. O. Zariski, P. Samuel, *Commutative Algebra*, vol. II. The University Series in Higher Mathematics (D. Van Nostrand, Princeton, 1960)

Index

© The Authors 2015

E. Guardo, A. Van Tuyl, *Arithmetically Cohen-Macaulay Sets of Points in* $\mathbb{P}^1 \times \mathbb{P}^1$,
SpringerBriefs in Mathematics, DOI 10.1007/978-3-319-24166-1

Printed in the United States
By Bookmasters